Intelligent Speech Signal Processing

Intelligent Speech Signal Processing

Edited by

Nilanjan Dey
Department of Information Technology,
Techno India College of Technology,
Kolkata, India

Academic Press is an imprint of Elsevier
125 London Wall, London EC2Y 5AS, United Kingdom
525 B Street, Suite 1650, San Diego, CA 92101, United States
50 Hampshire Street, 5th Floor, Cambridge, MA 02139, United States
The Boulevard, Langford Lane, Kidlington, Oxford OX5 1GB, United Kingdom

Library of Congress Cataloging-in-Publication Data
A catalog record for this book is available from the Library of Congress

British Library Cataloguing-in-Publication Data
A catalogue record for this book is available from the British Library

ISBN: 978-0-12-818130-0

Publisher: Mara Conner
Acquisition Editor: Mara Conner
Editorial Project Manager: Thomas Van Der Ploeg
Production Project Manager: Punithavathy Govindaradjane
Cover Designer: Miles Hitchen

Typeset by SPi Global, India

Contents

Contributors

Numbers in parenthesis indicate the pages on which the authors' contributions begin.

Rajesh Kumar Aggarwal (5), National Institute of Technology Kurukshetra, Kurukshetra, India

Mazid Alam (175), Department of CSE, Kaziranga University, Jorhat, India

David Brown (39), Sat-Com (PTY) Ltd., Windhoek, Namibia

Himanish Shekhar Das (81), Department of Computer Science and Engineering, National Institute of Technology Silchar, Silchar, India

Anilesh Dey (139), Department of Electronics and Communication Engineering, Narula Institute of Technology, Kolkata, India

Smita Dey (175), Department of CSE, Kaziranga University, Jorhat, India

Juan C. Gómez (55), Laboratory for System Dynamics and Signal Processing, Universidad Nacional de Rosario, CIFASIS-CONICET, Rosario, Argentina

Dharm Singh Jat (101), Namibia University of Science and Technology, Windhoek, Namibia

S. Jothilakshmi (113), Department of Information Technology, Annamalai University, Chidambaram, India

Biswajit Karan (153), Department of Electronics and Communication Engineering, Birla Institute of Technology, Mesra, Ranchi, India

Anton Sokamato Limbo (101), Namibia University of Science and Technology, Windhoek, Namibia

Kartik Mahto (153), Department of Electronics and Communication Engineering, Birla Institute of Technology, Mesra, Ranchi, India

Rajesh Kumar Muthu (39), Vellore Institute of Technology, Vellore, India

A. NithyaKalyani (113), Department of Computer Science and Engineering, Annamalai University, Chidambaram, India

Vishal Passricha (5), National Institute of Technology Kurukshetra, Kurukshetra, India

Pinki Roy (81), Department of Computer Science and Engineering, National Institute of Technology Silchar, Silchar, India

Gonzalo D. Sad (55), Laboratory for System Dynamics and Signal Processing, Universidad Nacional de Rosario, CIFASIS-CONICET, Rosario, Argentina

Sajal Saha (175), Department of CSE, Kaziranga University, Jorhat, India

Sitanshu Sekhar Sahu (153), Department of Electronics and Communication Engineering, Birla Institute of Technology, Mesra, Ranchi, India

K.C. Santosh (1), Department of Computer Science, The University of South Dakota, Vermillion, SD, United States

Sudhangshu Sarkar (139), Department of Electrical Engineering, Narula Institute of Technology, Kolkata, India

Charu Singh (39, 101), Sat-Com (PTY) Ltd., Windhoek, Namibia

Lucas D. Terissi (55), Laboratory for System Dynamics and Signal Processing, Universidad Nacional de Rosario, CIFASIS-CONICET, Rosario, Argentina

Maarten Venter (39), Sat-Com (PTY) Ltd., Windhoek, Namibia

About the Editor

Nilanjan Dey is an assistant professor in the Department of Information Technology at Techno India College of Technology, Kolkata. He completed his PhD from Jadavpur University in 2015. He is a visiting fellow at the Wearables Computing Laboratory, Department of Biomedical Engineering University of Reading, UK; visiting professor at the College of Information and Engineering, Wenzhou Medical University, China, and Duy Tan University, Vietnam. He has held the honorary position of visiting scientist at Global Biomedical Technologies Inc., CA, USA (2012–15).

He is the editor-in-chief of the *International Journal of Ambient Computing and Intelligence* (IGI Global), series co-editor of *Springer Tracts in Nature-Inspired Computing* (Springer) and *Advances in Ubiquitous Sensing Applications for Healthcare* (AUSAH; Elsevier), and series editor for *Intelligent Signal Processing and Data Analysis* (CRC Press). He has authored/edited more than 40 books for Elsevier, Wiley, CRC Press, and Springer, and published more than 350 research articles. His primary research interests include medical imaging, machine learning, bioinspired computing, data mining, and related fields. He is a life member of the Institute of Engineers (India).

Preface

Intelligent speech signal processing methods have increasingly replaced the conventional analog signal processing methods in several applications, including speech analysis and processing, telecommunications, and tracking. These intelligent speech signal processing approaches support different areas in a variety of everyday problems, multimedia communications, industrial automation, and biometrics. Incorporating different signal processing approaches, such as signal analysis using an analytical signal description, can be combined for efficient speech detection. In intelligent systems, pattern recognition and machine learning methods are vital tools for reasoning under uncertainty. They help to extract significant information from massive data in an automated fashion using statistical and computational methods. This domain is related to probability, statistics, optimization methods, and control theory. The focus is on providing solutions for tasks at which intelligence is inevitably essential. Application domains include computer vision, speech processing, natural language processing, man–machine interfaces, expert systems, and robotics, etc. Typically, there are general attributes that should be included in the intelligent signal processing system, namely nonlinearity, adaptively, and robustness. A speech signal processing device that operates in a nonstationary environment can be considered intelligent once it is able to explore the information content of its input in an efficient mode and at all times.

This book highlights researchers from machine learning, data analysis, data management, and speech processing provider fields. The authors sought trends and techniques in intelligent speech signal processing and data analysis to spotlight scientific breakthroughs in applied applications. The book includes 10 chapters. In Chapter 1, Santosh focuses on speech recognition/processing/synthesis in healthcare. He provides detailed information about how speech synthesis impacts healthcare and how it also impacts its business model. In Chapter 2, Passricha and Aggarwal discuss end-to-end acoustic modeling using the Conventional Neural Network (CNN) to establish the relationship between the raw speech signal and phones in a data-driven manner. This system has superior performance compared to the traditional cepstral feature-based systems, however, it requires a large number of parameters. In Chapter 3, Singh et al. propose a real-time DSP-based system for voice activity detection and background noise reduction. In Chapter 4, Sad et al. introduce a novel system to disambiguate conflict classification results in audio visual speech recognition (AVSR) applications. The performance of the proposed recognition system is evaluated on three publicly available audio-visual datasets, using the generative Hidden Markov Model, and three discriminative techniques, viz. random forests, support vector machines, and adaptive boosting. In Chapter 5, Das and Roy provide in-depth concepts of various Deep Learning techniques for spoken language identification, including their advantages and limitations. In Chapter 6, Jat et al. suggest a conceptual system design to enable people to automate processes in the home by using voice commands. In Chapter 7, NithyaKalyani and Jothilakshmi discuss several approaches for extractive and abstractive speech summarization, and they investigate speech summarization in the Indian language. Additionally, the chapter analyzes various speech recognition techniques and their performance on recognizing Tamil speech data. In Chapter 8, Sarkar and Dey introduce the dynamics of emotional speech signals using recurrence analysis. In Chapter 9, Karan et al. introduced nonconventional techniques for speech processing that overcame the problem of short-time processing of the speech signal. In Chapter 10, Saha et al. discuss the artificially intelligent customized voice response system design using speech synthesis markup language. This chapter introduces a low-cost artificially intelligent voice response system driven by the Amazon Web Server on an IoT cloud platform and Raspberry Pi.

This book supports and enhances the utilization of speech analytics in several systems and real-world activities. It provides a well-standing forum to discuss the characteristics of the intelligent speech signal processing systems in different domains. The book is proposed for professionals, scientists, and engineers who are involved in the new techniques of intelligent speech signal processing methods and systems.

Chapter 1

Speech Processing in Healthcare: Can We Integrate?

K.C. Santosh
Department of Computer Science, The University of South Dakota, Vermillion, SD, United States

Speech recognition—also known as name voice recognition—refers to the translation from speech into words in a machine-readable format [1–3].

Speech processing has been considered for various purposes in the domain, for example, signal processing, pattern recognition, and machine learning [3]. Starting with the improvement of customer service, as well as the role of hospital care in combating crime, among other purposes, we have found that speech recognition has increased its global market from $104.4 billion in 2016 to an estimated $184.9 billion in 2021 (source: https://www.news-medical.net/whitepaper/20170821/Speech-Recognition-in-Healthcare-a-Significant-Improvement-or-Severe-Headache.aspx). This is not a new trend; for cases in which different languages are needed, speech-to-text conversion is an example that has been widely used. The opposite holds true as well [4, 5]. For example, can we process or reuse speech data that occurred during a telephone conversation a few years previously, in which a client claimed that fraud happened on his or her credit card? Yes, this is possible. Beside other sources of data, speech can be taken as an authentic component to describe an event or scene wherein emotions can be analyzed [6–8].

Examples exist showing how speech analysis can be integrated into healthcare. In Fig. 1.1, a complete healthcare automated scenario has been created, which can be summarized as follows:

> *A patient visits clinical center (hospital), where he/she gets X-rayed, provides sensor-based data (external and internal), and receives (handwritten and machine-printed) prescription(s) and report(s) from the specialist. In these events, a patient and other staff (including the specialists) have gone through different levels of conversation, and, if recorded, they will be able to integrate these with signal processing, pattern recognition, image processing, and machine learning.*

In the aforementioned healthcare project for instance, it would be convenient to combine speech and signal processing tools and techniques with image

Intelligent Speech Signal Processing. https://doi.org/10.1016/B978-0-12-818130-0.00001-5

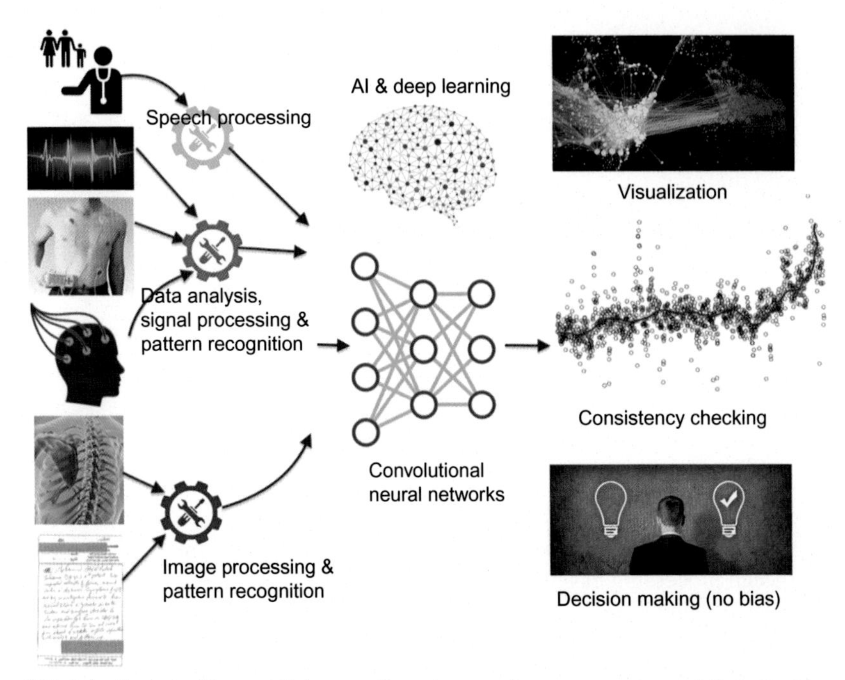

FIG. 1.1 Smart healthcare and the use of speech processing: can we get more information?

analysis-based tools and techniques [9–12]. More specifically, it is important to note that doctors can predict or guess about the presence of tuberculosis, for instance, based on verbal communication (quoted answers to questions, such as "do you sleep well?" and "how are your eating habits?") before they start the X-ray screening procedure. If this is the case, in the complete project (outlined earlier), speech processing can help. This means that we would be able to come up with the complete information so that further processing can be performed.

It is important to note, for instance, that speech and voice cues (before and after the doctor's visit) can help one understand the patient's willingness to continue with treatment. Speech and voice can definitely convey emotions over time. Further, pain can simply be read by speech/voice level. We can also automate and check the trends related to how doctors and other staff members behave toward patients. Can speech be a component in helping to find consistency that is evident in other sources of data, as shown in Fig. 1.1? The use of a (proposed) convolutional neural network, as in Fig. 1.1, helps explain the fact that a machine, unlike a human expert, for instance, can make a decision without the bias possible in human choices. Also, visualization is possible that can show how different sources of data are connected. We consider artificial intelligence (AI) and machine learning (ML) tools for automation since data have to be collected over time. Analyzing big data is extremely important since it is not possible manually because humans are more error-prone and human analysis is costlier.

As mentioned earlier, local languages other than English can be considered in healthcare. In one work [13], authors reported the use of the Tamil language, in addition to English, to estimate heartbeat from speech/voice. A few more can be cited, showing how local, regional languages, such as German [14], Malay [15], and Slovenian [16], have helped speech technology progress. In another context [17], emergency medical care often depends on how quickly and accurately field medical personnel can access a patient's background information and document their assessment and treatment of the patient. What if we could automate speech/voice recognition tools in the field? It is clear that research scientists should come up with precise tools that people can trust. Analyzing speech/voice concurrently with background music or other noise is important. Other works can be referenced for more detailed information [18–21].

As data change and increase, machine learning could help to automate the data retrieval and recording system. The use of the extreme-learning voice activity detection is prominent in the field [22]. As we go further, for example with real-time speech/voice recognition/classification, active learning should be considered since scientists found that learning over time is vital [23].

In general, Fig. 1.1 shows how important different sources of data and speech are as components in things such as sensor-based and image data (X-ray and/or reports that are handwritten or machine printed).

References

[1] J. Flanagan, L. Rabiner (Eds.), Speech Synthesis, Dowden, Hutchinson & Ross, Inc., Pennsylvania, 1973.

[2] J. Flanagan, Speech Analysis, Synthesis, and Perception, Springer-Verlag, Berlin-Heidelberg-New York, 1972.

[3] C. Xian-Yi, P. Yan, Review of modern speech synthesis, in: W. Hu (Ed.), Electronics and Signal Processing. Lecture Notes in Electrical Engineering, vol. 97, Springer, Berlin, Heidelberg, 2011.

[4] A. Iida, N. Campbell, Speech database design for a concatenative text-to-speech synthesis system for individuals with communication disorders, Int. J. Speech Technol. 6 (4) (2003) 379–392.

[5] R. Bossemeyer, M. Hardzinski, Talking call waiting: an application of text-to-speech, Int. J. Speech Technol. 4 (1) (2001) 7–17.

[6] K. Sailunaz, M. Dhaliwal, J. Rokne, R. Alhajj, Emotion detection from text and speech: a survey, Soc. Netw. Anal. Min. 8 (28) (2018).

[7] A. Revathi, C. Jeyalakshmi, Emotions recognition: different sets of features and models, Int. J. Speech Technol. (2018).

[8] M. Swain, A. Routray, P. Kabisatpathy, Databases, features and classifiers for speech emotion recognition: a review, Int. J. Speech Technol. 21 (1) (2018) 93–120.

[9] K.C. Santosh, S. Antani, Automated chest X-ray screening: can lung region symmetry help detect pulmonary abnormalities? IEEE Trans. Med. Imaging 37 (5) (2018) 1168–1177.

[10] S. Vajda, A. Karagyris, S. Jaeger, K.C. Santosh, S. Candemir, Z. Xue, S. Antani, G. Thoma, Feature selection for automatic tuberculosis screening in frontal chest radiographs, J. Med. Syst. 42 (8) (2018) 146.
[11] A. Karargyris, J. Siegelman, D. Tzortzis, S. Jaeger, S. Candemir, Z. Xue, K.C. Santosh, S. Vajda, S.K. Antani, L.R. Folio, R. George, Thoma: combination of texture and shape features to detect tuberculosis in digital chest X-rays, Int. J. Comput. Assist. Radiol. Surg. 11 (1) (2016) 99–106.
[12] K.C. Santosh, S. Vajda, S.K. Antani, G.R. Thoma, Edge map analysis in chest X-rays for automatic abnormality screening, Int. J. Comput. Assist. Radiol. Surg. 11 (9) (2016) 1637–1646.
[13] A. Milton, K.A. Monsely, Tamil and English speech database for heartbeat estimation, Int. J. Speech Technol. (2018).
[14] The German text-to-speech synthesis system MARY: a tool for research, development and teaching, Int. J. Speech Technol. 6 (4) (2003) 365–377.
[15] Y.A. El-Imam, Z.M. Don, Text-to-speech conversion of standard Malay, Int. J. Speech Technol. 3 (2) (2000) 129–146.
[16] T. Šef, M. Gams, SPEAKER (GOVOREC): a complete Slovenian text-to-speech system, Int. J. Speech Technol. 6 (3) (2003) 277–287.
[17] G. Thomas, Holzman: speech-audio interface for medical information management in field environments, Int. J. Speech Technol. 4 (3–4) (2001) 209–226.
[18] B.K. Khonglah, S.R.M. Prasanna, Clean speech/speech with background music classification using HNGD spectrum, Int. J. Speech Technol. 20 (4) (2017) 1023–1036.
[19] B.K. Khonglah, S.M. Prasanna, Speech/music classification using speech-specific features, Digital Signal Process. 48 (2016) 71–83.
[20] E. Scheirer, M. Slaney, Construction and evaluation of a robust multifeature speech/music discriminator, in: Proceedings of the IEEE International Conference on Acoustics, Speech, and Signal Processing, Vol. 2, 1997, pp. 1331–1334.
[21] T. Zhang, C.J. Kuo, Audio content analysis for online audiovisual data segmentation and classification, IEEE Trans. Speech Audio Process. 9 (4) (2001) 441–457.
[22] H. Mukherjee, S.M. Obaidullah, K.C. Santosh, S. Phadikar, K. Roy, Line spectral frequency-based features and extreme learning machine for voice activity detection from audio signal, Int. J. Speech Technol. (2018).
[23] M.-R. Bouguelia, S. Nowaczyk, K.C. Santosh, A. Verikas, Active learning in the domain: agreeing to disagree: active learning with noisy labels without crowdsourcing, Int. J. Mach. Learn. Cybern. (2018).

Chapter 2

End-to-End Acoustic Modeling Using Convolutional Neural Networks

Vishal Passricha and Rajesh Kumar Aggarwal
National Institute of Technology Kurukshetra, Kurukshetra, India

2.1 Introduction

An automatic speech recognition (ASR) system has two important tasks—phoneme recognition and whole word decoding. In ASR, the relationship between the speech signal and phones are established in three steps [1]. First, useful features are extracted from the speech signal on the basis of prior knowledge. This phase is known as information selection, dimensionality reduction, or feature extraction phase. In this, the dimensionality of the speech signal is reduced by selecting the information based on task-specific knowledge. Highly specialized features like MFCC (Mel-frequency cepstral coefficient) [2] and PLP (perceptual linear prediction) [3] are the preferred choices in traditional ASR systems. Next, generative models like Gaussian mixture model (GMM), hidden Markov model (HMM), or discriminative models like artificial neural networks (ANNs), deep neural networks (DNNs), and convolutional neural networks (CNNs) estimate the likelihood of each phoneme. This step is known as acoustic modeling. Finally, word sequence is recognized using dynamic programming technique. This step is decision making, it integrates the acoustic model, lexical and language model to decode the text utterance. Deep learning systems can map the acoustic features into the spoken phonemes directly. A sequence of the phoneme is easily generated from the frames using frame level classification.

Another side, end-to-end systems perform acoustic frames to phone mapping in one step only. End-to-end training means all the modules are learned simultaneously. Advanced deep learning methods facilitate the training of the ASR system in an end-to-end manner. They can also train ASR system directly with raw signals, that is, without hand-crafted features. Therefore, the ASR paradigm is shifting from cepstral features like MFCC, PLP to

Intelligent Speech Signal Processing. https://doi.org/10.1016/B978-0-12-818130-0.00002-7

discriminative features learned directly from raw speech. The end-to-end model may take a raw speech signal as input and generates phoneme class conditional probabilities as output. The three major types of end-to-end architectures for ASR are attention-based method, connectionist temporal classification (CTC), and CNN-based direct raw speech model. End-to-end trained systems are gaining popularity in speech signal to word sequence [4–9].

Attention-based models directly transcribe the speech into phonemes by jointly training an acoustic model, language model, and alignment mechanism. Attention-based encoder-decoder uses the recurrent neural network (RNN) to perform sequence-to-sequence mapping without any predefined alignment. In this model, the input sequence is first transformed into a fixed-length vector representation and then the decoder maps this fixed-length vector into the output sequence. Attention-based encoder-decoder is capable of learning the mapping between variable-length input and output sequences. Chorowski and Jaitly [10] proposed speaker independent sequence-to-sequence model that achieved 10.6% word error rate (WER) without separate language models and 6.7% WER with a trigram language model for *Wall Street Journal* dataset. In attention-based systems, the alignment between the acoustic frames and recognized symbols is performed by the attention mechanism, whereas the CTC model uses conditional independence assumptions to solve sequential problems by dynamic programming efficiently. Attention model has shown high performance over the CTC approach because it uses the history of the target character without any conditional independence assumptions. Soltau et al. [11] performed context-dependent phoneme recognition by training the CTC-based model on YouTube video caption task. Sequence-to-sequence models lack behind 13%–35% than baseline systems. Graves et al. [12] trained the end-to-end model on CTC criterion without applying a frame-level alignment. A sequence-to-sequence model simplified the ASR problem by learning and optimizing the neural network for an acoustic model, pronunciation model, and language model [5, 7, 9, 13]. These models also work as multi-dialects systems because they are jointly modeled across dialects. Li and others [14] trained a single sequence to sequence model for multi-dialect speech recognition and achieved a similar performance as another sequence-to-sequence model for single dialect tasks.

Another side, the CNN-based acoustic model is proposed by Palaz et al. [15–17] which processes the raw speech directly as input. This model consists of two stages: Feature learning stage, that is, several convolutional layers and classifier stage, that is, fully connected layers. Both the stages are learned jointly by minimizing a cost function based on relative entropy. In this model, the features are extracted by the filters at the first convolutional layer and processed between the first and second convolutional layer. In the classifier stage, learned features are classified by fully connected layers and a softmax layer, that is, model the relation between features and phoneme. This approach claims comparable or better performance than traditional cepstral feature-based system followed by ANN trained for phoneme recognition on TIMIT dataset.

CNN has three advanced features over DNN i.e., convolutional filters, pooling, and weight sharing. Pooling operator is applied to extract lower & higher level features. The key features of pooling are achieving compact representation and more robustness to noise and clutter. Various pooling strategies are found obscure by several compounding factors. Various pooling strategies like max pooling, average pooling, stochastic pooling, L_p pooling, etc. have their own advantages and disadvantages. It is always critical to understand and select the best pooling technique. In this chapter, we compare various pooling strategies for end-to-end CNN-based direct raw speech model. We further explore the effects of nonlinearity of fully connected layers on the recognition rate.

This chapter is organized as follows: In Section 2.2, the work performed in the field of ASR is discussed with the name of related work. Section 2.3 covers the various architectures of ASR. Section 2.4 presents the brief introduction about CNN. Section 2.5 explains the CNN-based direct raw speech recognition model. In Section 2.6, the available experiment and their results are shown. Finally, Section 2.7 concludes this chapter with a brief discussion.

2.2 Related Work

A traditional ASR system leveraged the GMM/HMM paradigm for acoustic modeling. GMM efficiently processes the vectors of input features and estimates emission probabilities for each HMM state. HMM efficiently normalizes the temporal variability present in the speech signal. The combination of HMM and the language model is used to estimate the most likely sequence of phones. The discriminative objective functions, that is, maximum mutual information, minimum classification error, minimum phone error, and minimum word error (MWE) are used to improve the recognition rate of the system by the discriminative training methods [18]. However, GMM has a shortcoming because it shows an inability to model the data that lie on or near a nonlinear manifold in the dataspace [19]. Therefore its performance degrades in the presence of noise. On the other hand, ANNs can learn much better models of data that lay in the noisy conditions. DNNs as acoustic models tremendously improved the performance of ASR systems [19–21]. Generally, discriminative power of DNNs is used for phoneme recognition and for decoding task, HMM is the preferred choice. DNNs have many hidden layers with a large number of nonlinear units and a very large output layer. The benefit of this large output layer is that it accommodates a large number of HMM states that arise when each phone is modeled by a number of different “triphone” HMMs, which take into account the phones on either side. DNN architecture has densely connected layers. Therefore such architectures are more prone to overfitting. Secondly, features that have the local correlations become difficult to learn by such architectures. On the other hand, in [22], speech frames are classified into clustered context-dependent states using DNNs. In [23, 24], GMM-free DNN training process is proposed by the researchers. However, GMM-free process demands iterative procedures like decision trees and generating forced alignments. DNN-based acoustic models are gaining popularity in large vocabulary speech recognition tasks [19], but components like HMM and an n-gram language model are the same as in their predecessors.

GMM- or DNN-based ASR systems perform the task in three steps: feature extraction, classification, and decoding. It is shown in Fig. 2.1. Firstly, the short-term signal s_t are processed at time 't' to extract the features x_t. These features are provided as input to GMM or DNN acoustic model which estimates the class conditional probabilities $P_e(i|x_i)$ for each phone class $i \in \{1,\dots,I\}$. The emission probabilities are as follows:

$$p_e(x_t|\,i) \propto \frac{p(x_t|\,i)}{p(x_t)} = \frac{p(i|\,x_t)}{p(i)} \quad \forall_i \in i,\dots,I \tag{2.1}$$

The prior class probability $p(i)$ is computed by counting on the training set.

DNN is a feed-forward NN containing multiple hidden layers with a large number of hidden units. DNNs are trained using the back-propagation methods, then they are discriminatively fine-tuned for reducing the gap between desired output and actual output. DNN/HMM-based hybrid systems are the effective models that use a triphone HMM model and an n-gram language model [19, 25]. Traditional DNN/HMM hybrid systems have several independent components that are trained separately like an acoustic model, pronunciation model, and language model. In the hybrid model, the speech recognition task is factorized into several independent subtasks. Each subtask is independently handled by a separate module, which simplifies the objective. The classification task is much simpler in HMM-based models as compared to classifying the set of variable length sequences directly. This is because sequence classification with variable length requires sequence padding of data and some advanced models like long short-term memory (LSTM), bidirectional long short-term memory (BLSTM), etc. Fig. 2.2 shows the hybrid DNN/HMM phoneme recognition model.

On the other side, researchers proposed end-to-end ASR systems that directly map the speech into labels without any intermediate components.

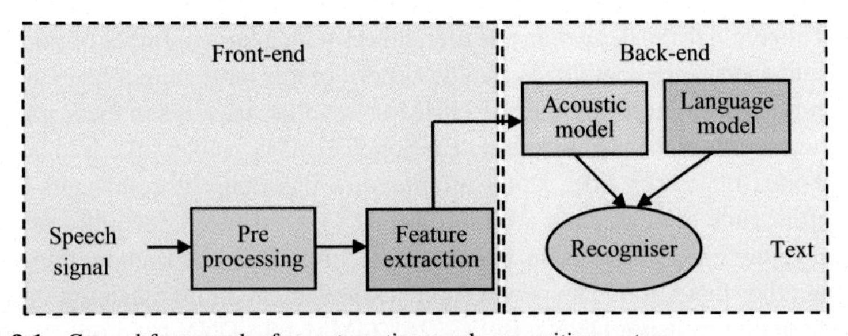

FIG. 2.1 General framework of an automatic speech recognition system.

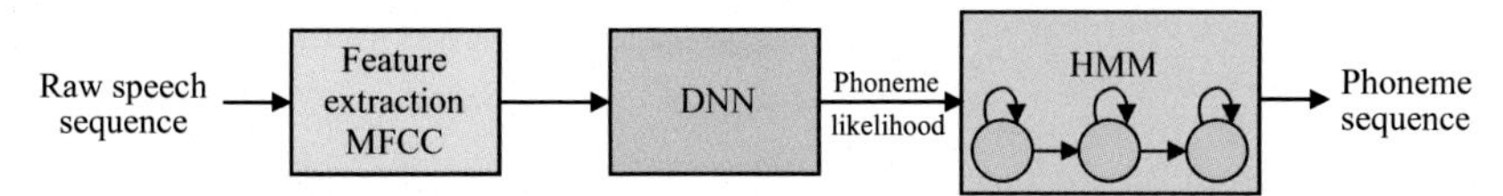

FIG. 2.2 Hybrid DNN/HMM phoneme recognition.

As an advancement in deep learning, it became possible to train the system in an end-to-end fashion. The high success rate of deep learning methods in vision task motivates the researchers to focus on the classifier step for speech recognition. Such architectures are called deep because they are composed of many layers as compared to classical shallow systems. The main goal of an end-to-end ASR system is to simplify the conventional module-based ASR system into a single Deep Learning framework. In earlier systems, divide and conquer approaches are used to optimize each step independently whereas Deep Learning approaches have a single architecture that leads to the more optimal system. End-to-end speech recognition systems directly map the speech into text without requiring predefined alignment between acoustic frame and characters [5–7, 9, 26–30]. These systems are generally divided into three broad categories: Attention-based model [6, 7, 28, 29], CTC [5, 12, 26, 27], and CNN-based direct raw speech method [15–17, 31]. All these models can address the problem of variable-length input and output sequences.

Attention-based models are gaining popularity in a variety of tasks like handwriting synthesis [32], machine translation [33], and visual object classification [34]. Attention-based models directly map the acoustic frame into character sequences. However, this model differs from other machine translation tasks by requesting much longer input sequences. This model generates a character based on the inputs and history of the target character. The attention-based models use encoder-decoder architecture to perform the sequence mapping from speech feature sequences to text as shown in Fig. 2.3. Its extension, that is, attention-based recurrent networks have also been successfully applied to speech recognition. In the noisy environment, these models' results are poor because noise easily corrupts the estimated alignment. Another issue with this model is that it is hard to train from scratch due to misalignment on longer input sequences. Sequence-to-sequence networks have also achieved many breakthroughs in speech recognition [6, 7, 29]. They can be divided into three modules: an encoding module that transforms sequences; an attention module that estimates the alignment between the hidden vector and targets; and the decoding module that generates the output sequence. It replaces the complex data processing pipelines with a single neural network trained in end-to-end fashion. It is directly trained by discriminative training methods to maximize the probability of observing desired outputs conditioned on the inputs. The discriminative training is a different way of training that raises the performance of the system by focusing on most informative features. It also overcomes the risk of overfitting, so it

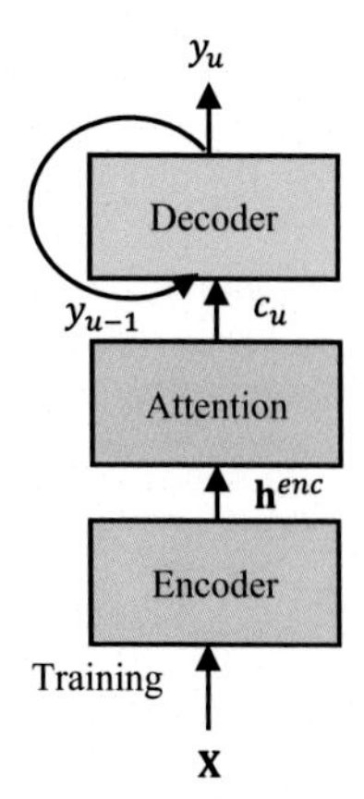

FIG. 2.3 Listen, attend, and spell model including an encoder, attention, and decoder.**X** represents input sequence of filterbank spectra. $\mathbf{h}^{enc}$ represents the high level feature representation. c_u represents the context vector. y_{u-1} represents the previously emitted character. y_u represents the next emitted character and Y be the output sequence of characters.

is currently gaining high popularity. Therefore to develop a successful sequence-to-sequence model, the understanding and preventing limitations are required.

End-to-end trainable speech recognition systems are an important application of attention-based models. The decoder network computes a matching score between hidden states generated by the acoustic encoder network at each input time. It processes its hidden states to form a temporal alignment distribution. This matching score is used to estimate the corresponding encoder states. The difficulty of an attention-based mechanism in speech recognition is that the feature inputs and corresponding letter outputs generally proceed in the same order with only small deviations within word. However, the different length of input and output sequences make it more difficult to track the alignment. The advantage of an attention-based mechanism is that any conditional independence assumptions (Markov assumption) are not required in this mechanism. Attention-based approach replaces the HMM with RNN to perform the sequence prediction (Fig. 2.4). The attention mechanism automatically learns alignment between the input features and desired character sequence.

CTC techniques infer the speech-label alignment automatically. CTC [12] was developed for decoding the language. Hannun et al. [26] used it for decoding purposes in Baidu's deep speech network. CTC uses dynamic programming [5] for efficient computation of a strictly monotonic alignment. However, graph-based decoding and a language model are required for it. CTC approaches use RNN for feature extraction [33]. Graves et al. [35] used its objective function in BLSTM system. This model successfully arranges all possible alignments between input and output sequences during model training.

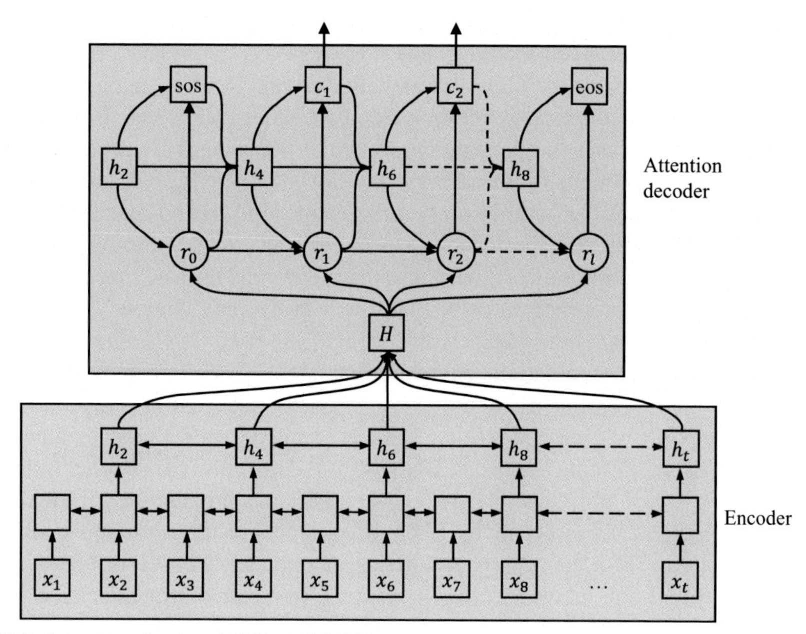

FIG. 2.4 Attention-based ASR model [33].

Two different versions of beam search are adopted by [5, 36] for decoding CTC models. Fig. 2.5 shows the working architecture of the CTC model. In this, noisy and noninformative frames are discarded by the introduction of the blank label, which results in the optimal output sequence. CTC uses intermediate label representation to identify the blank labels, that is, no output labels. CTC-based NN model shows a high recognition rate for both phoneme recognition [4] and large vocabulary continuous speech recognition (LVCSR) [5, 36]. CTC trained neural network with language model offers excellent results [26].

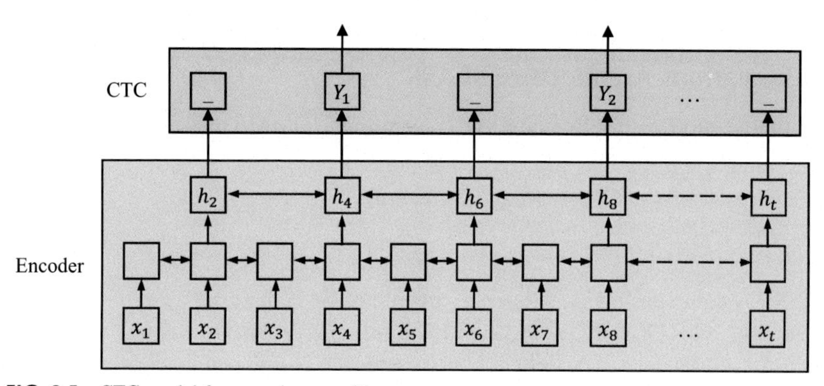

FIG. 2.5 CTC model for speech recognition.

In Fig. 2.5, $\{x_1, x_2, \ldots, x_t\}$ represents our input sequence, $\{h_2, h_4, \ldots, h_t\}$ represents the high-level features, and $\{Y_1, Y_2, \ldots\}$ represents the letter sequence generated by the CTC decoder. End-to-end ASR systems perform well and achieve good results, yet they face two major challenges. First, how to incorporate lexicons and language models into decoding. However, [5, 36] has incorporated lexicons for searching paths. Second, there is no shared experimental platform for the purpose of the benchmark. End-to-end systems differ from the traditional systems in both aspects, model architecture as well as decoding methods. Some efforts were also made to model the raw speech signal with little or no preprocessing [37]. Earlier, Jaitly and Hinton [37] used restricted Boltzmann machines, Tüske et al. [38] used DNN and Sainath et al. [39] used the combination of convolutional, long short-term memory with fully connected DNN to process the raw waveforms. Although, results did not meet the previously available results. The NN architectures with local connections and shared weights are called CNN [40]. Palaz et al. [16] showed in his study that CNN could calculate the class conditional probabilities from the raw speech signal as direct input. Therefore CNNs are the preferred choice to learn features from the raw speech. There are two stages of learned features process. Features are learned by the filters at the first convolutional layer, then learned features are modeled by second and higher level convolutional layers. An end-to-end phoneme sequence recognizer directly processes the raw speech signal as input and produces a phoneme sequence. This end-to-end system is composed of two parts: CNNs and conditional random field (CRF). CNN is used to perform the feature learning and classification, and CRFs are used for the decoding stage. CRF, ANN, multilayer perceptron etc. have been successfully used as decoder. The results on TIMIT phone recognition task also confirm that the system effectively learns the features from raw speech and performs better than traditional systems that take cepstral features as input [41]. This model also produced good results for LVCSR [17]. Table 2.1 shows the comparison of different end-to-end speech recognition model in the form of WER.

2.3 Various Architecture of ASR

In this section, a brief review on conventional GMM/DNN ASR, attention-based end-to-end ASR, and CTC is given.

2.3.1 GMM/DNN

ASR system performs sequence mapping of T-length speech sequence features, $X = \{X_t \in \mathbb{R}^D \mid t = 1, \ldots, T\}$, into an N-length word sequence, $W = \{w_n \in \upsilon \mid n = 1, \ldots, N\}$ where X_t represents the D-dimensional speech feature vector at frame t, and w_n represents the word at position n in the vocabulary υ.

TABLE 2.1 WERs (%) on Various Test Sets [13]

Model	Clean		Noisy		Numeric Values Only
	Isolated Words	*Continuous Speech*	*Isolated Words*	*Continuous Speech*	
Baseline unidirectional context-dependent phoneme	6.4	9.9	8.7	14.6	11.4
Baseline bidirectional context-dependent phoneme	5.4	8.6	6.9	–	11.4
End-to-end systems					
CTC grapheme	39.4	53.4	–	–	–
RNN transducer	6.6	12.8	8.5	22.0	9.9
RNN transducer with attention	6.5	12.5	8.4	21.5	9.7
Attention 1-layer decoder	6.6	11.7	8.7	20.6	9.0
Attention 2-layer decoder	6.3	11.2	8.1	19.7	8.7

The ASR problem is formulated within the Bayesian framework. In this method, an utterance is represented by some sequence of acoustic feature vector X, derived from the underlying sequence of words W, and the recognition system needs to find the most likely word sequence, given below [42]:

$$\widehat{W} = \arg\max_{w} p(W|X) \tag{2.2}$$

In Eq. (2.2), the argument of $p(W|X)$, that is, the word sequence W is found which shows the maximum probability for given feature vector, X. Using Bayes' rule, it can be written as:

$$\widehat{W} = \arg\max_{w} \frac{p(X|W)p(W)}{p(X)} \tag{2.3}$$

In Eq. (2.3), the denominator $p(X)$ is ignored as it is constant with respect to W. Therefore,

$$\widehat{W} = \arg\max_{w} p(X|W)p(W) \tag{2.4}$$

where $p(X|W)$ represents the likelihood of the feature vector X and it is evaluated with the help of an acoustic model. $p(W)$ represents the prior knowledge, that is, a priori probability about the sequence of words W and it is determined by the language model. However, current ASR systems are based on a hybrid HMM/DNN [43], that is also calculated using Bayes' theorem and introduces the HMM state sequence S, to factorize $p(W|X)$ into the following three distributions:

$$\arg\max_{w \in v^*} p(W|X) \tag{2.5}$$

$$= \arg\max_{w \in v^*} \sum_{S} p(X|S,W)p(S|W)p(W) \tag{2.6}$$

$$\approx \arg\max_{w \in v^*} \sum_{S} p(X|S),p(S|W)p(W) \tag{2.7}$$

where $p(X|S)$, $p(S|W)$, and $p(W)$ represent acoustic, lexicon, and language models respectively. v^* represents the all possible word sequence. Eq. (2.6) is changed into Eq. (2.7) by a conditional independence assumption. *Note:* It is a reasonable assumption to simplify the dependency of the acoustic model.

(a) *Acoustic models* $p(X|S)$: $p(X|S)$ can be further factorized using a probabilistic chain rule and Markov assumption as follows:

$$p(X|S) = \prod_{t=1}^{T} p(x_t|x_1,\ldots,x_{t-1},S) \tag{2.8}$$

$$\approx \prod_{t=1}^{T} p(x_t|s_t) \propto \prod_{t=1}^{T} \frac{p(s_t|x_t)}{p(s_t)} \tag{2.9}$$

In Eq. (2.9), framewise likelihood function $p(x_t|s_t)$ is changed into the framewise posterior distribution $\frac{p(s_t|x_t)}{p(s_t)}$, which is computed using DNN classifiers by pseudo-likelihood trick [43]. In Eq. (2.9), the Markov assumption is too strong. Therefore the contexts of input and hidden states are not considered. This issue can be resolved using either the RNNs or DNNs with long context features. A framewise state alignment is required to train the framewise posterior, which is offered by an HMM/GMM system.

(b) *Lexicon model* $p(S|W)$: $p(S|W)$ can be further factorized using a probabilistic chain rule and Markov assumption (first order) as follows:

$$p(S|W) = \prod_{t=1}^{T} p(s_t|s_1,\ldots,s_{t-1},W) \tag{2.10}$$

$$\approx \prod_{t=1}^{T} p(s_t | s_{t-1}, W) \tag{2.11}$$

An HMM state transition represents this probability. A pronunciation dictionary performs the conversion from w to HMM states through phoneme representation.

(c) *Language model* $p(W)$*:* Similarly, $p(W)$ can be factorized using a probabilistic chain rule and Markov assumption ($(m-1)$th order) as an m-gram model, that is,

$$p(W) = \prod_{n=1}^{N} p(w_n | w_1, \ldots, w_{n-1}) \tag{2.12}$$

$$\approx \prod_{n=1}^{N} p(w_n | w_{n-m-1}, \ldots, w_{n-1}) \tag{2.13}$$

The issue of Markov assumption is addressed using RNN language model (RNNLM) [44], but it increases the complexity of the decoding process. The combination of RNNLMs and m-gram language model is generally used and it works on a rescoring technique.

2.3.2 Attention Mechanism

The approaches based on attention mechanism does not make any Markov assumptions. It directly finds the posterior probability $p(C|X)$, on the basis of a probabilistic chain rule.

$$p(C|X) = \underbrace{\prod_{l=1}^{L} p(c_l | c_1, \ldots, c_{l-1}, X)}_{\triangleq p_{att}(C/X)} \tag{2.14}$$

where $p_{att}(C|X)$ represents an attention-based objective function. C represents the L-length letter sequence. $p(c_l | c_1, \ldots, c_{l-1}, X)$ is obtained by

$$\boldsymbol{h}_t = Encoder(X) \tag{2.15}$$

$$a_{lt} = \begin{cases} ContentAttention(\boldsymbol{q}_{l-1}, \boldsymbol{h}_t) \\ LocationAttention(\{a_{l-1}\}_{t=1}^{T}, \boldsymbol{q}_{l-1}, \boldsymbol{h}_t) \end{cases} \tag{2.16}$$

$$\boldsymbol{r}_l = \sum_{t=1}^{T} a_{lt} \boldsymbol{h}_t \tag{2.17}$$

$$p(c_l | c_1, \ldots, c_{l-1}, X) = Decoder(\boldsymbol{r}_l, \boldsymbol{q}_{l-1}, c_{l-1}) \tag{2.18}$$

Eq. (2.15) represents the encoder and Eq. (2.18) represents the decoder networks. a_{lt} represents the soft alignment of the hidden vector, $\boldsymbol{h}_t$. It is also known as attention weight. Here, $\boldsymbol{r}_l$ represents the weighted letter-wise hidden vector that is computed by weighted summation of hidden vectors. Content-based attention mechanism with or without convolutional features are represented by *ContentAttention*(·) and *LocationAttention*(·), respectively.

(a) *Encoder network:* The input speech feature sequence X is converted into a framewise hidden vector, $\boldsymbol{h}_t$ using Eq. (2.15). The preferred choice for an encoder network is BLSTM that is,

$$Encoder(X) \triangleq BLSTM_t(X) \tag{2.19}$$

It is to be noted that the computational complexity of the encoder network is reduced by subsampling the inputs [6, 7].

(b) *Content-based attention mechanism: ContentAttention*(·) is shown as

$$e_{lt} = g^T \tanh\left(Lin(\boldsymbol{q}_{l-1}) + LinB(\boldsymbol{h}_t)\right) \tag{2.20}$$

$$a_{lt} = Softmax\left(\{e_{lt}\}_{t=1}^{T}\right) \tag{2.21}$$

g represents a learnable parameter. $\{e_{lt}\}_{t=1}^{T}$ represents a T-dimensional vector. tanh(·) and *Lin*(·) represent the hyperbolic tangent activation function, and linear layer with learnable matrix parameters respectively.

(c) *Location-aware attention mechanism:* It is an extended version of content-based attention mechanism to deal with the location-aware attention (i.e., convolution). We know that $a_{l-1} = \{a_{l-1}\}_{t=1}^{T} = [a_{l-1,1}, a_{l-1,2}, \cdots, a_{l-1,T}]^T$. If $a_{l-1} = \{a_{l-1}\}_{t=1}^{T}$ is replaced in Eq. (2.16), then *LocationAttention*(·) is represented as follows:

$$\{\boldsymbol{f}_t\}_{t=1}^{T} = \mathcal{K} * \boldsymbol{a}_{l-1} \tag{2.22}$$

$$e_{lt} = g^T \tanh\left(Lin(\boldsymbol{q}_{l-1}) + Lin(\boldsymbol{h}_t) + LinB(\boldsymbol{f}_t)\right) \tag{2.23}$$

$$a_{lt} = softmax\left(\{e_t\}_{t=1}^{T}\right) \tag{2.24}$$

Here, * denotes 1-D convolution along the input feature axis, t, with the convolution parameter, $\mathcal{K}$, to produce the set of T features $\{f_t\}_{t=1}^{T}$. *LinB*(·) represents the linear layer with learnable matrix parameters with bias vector parameters.

(d) *Decoder network:* The decoder network is an RNN that is conditioned on previous output C_{l-1} and hidden vector $\boldsymbol{q}_{l-1}$. LSTM is the preferred choice of RNN that is represented as follows:

$$Decoder(\cdot) \triangleq softmax(LinB(LSTM_l(\cdot)) \tag{2.25}$$

$LSTM_l(\cdot)$ represents uniconditional LSTM that generates hidden vector $\boldsymbol{q}_l$ as output.

$$\boldsymbol{q}_l = LSTM_l(\boldsymbol{r}_l, \boldsymbol{q}_{l-1}, c_{l-1}) \tag{2.26}$$

$\boldsymbol{r}_l$ represents the concatenated vector of the letter-wise hidden vector, c_{l-1} represents the output of the previous layer which is taken as input.

(e) *Objective function:* The training objective of the attention model is computed from the sequence posterior

$$p_{att}(C|X) \approx \prod_{l=1}^{L} p\left(c_l | c_1^*, \ldots, c_{l-1}^*, X\right) \triangleq p_{att}^*(C|X) \tag{2.27}$$

where c_l^* represents the ground truth of the previous characters. The attention-based approach is a combination of letter-wise objectives based on multiclass classification with the conditional ground truth history $c_1^*, \ldots, c_{l-1}^*$ in each output l, and does not fully consider a sequence-level objective, as pointed out by [6].

2.3.3 Connectionist Temporal Classification

The CTC formulation is also based on Bayes' decision theory. CTC formulation uses letter sequence C with blank symbol to denote the letter boundary to handle the repetition of letter symbols. An augmented letter sequence

$$\begin{aligned} C' &= \{\langle b \rangle, c_1, \langle b \rangle, c_2, \langle b \rangle, \ldots, c_L, \langle b \rangle\} \\ &= \left\{c_l' \in \mathcal{U} \cup \{\langle b \rangle\} \,|\, l = 1, \ldots, 2L+1\right\} \end{aligned} \tag{2.28}$$

In C', c_l' represents letter symbol produced. $\mathcal{U}$ represents the set of distinct letters. l is the number between 1 and $2L+1$ that decides the produced symbol is blank or letter symbol. c_l' is always "$\langle b \rangle$" and letter when l is an odd and an even number, respectively. Similar to DNN/HMM model, framewise letter sequence with an additional blank symbol

$$Z = \{z_t \in \mathcal{U} \cup \{\langle b \rangle\} \,|\, t = 1, \ldots, T\} \tag{2.29}$$

is also introduced. The posterior distribution, $p(C|X)$, can be factorized as

$$p(C|X) = \sum_{Z} p(C|Z,X) p(Z|X) \tag{2.30}$$

$$\approx \sum_{Z} p(C|Z).p(Z|X) \tag{2.31}$$

Same as Eq. (2.7), CTC also uses Markov assumption, that is, $p(C|Z,X) \approx p(C|Z)$ to simplify the dependency of the CTC acoustic model, $p(Z|X)$, and CTC letter model, $p(C|Z)$.

(a) *CTC acoustic model:* Same as DNN/HMM acoustic model, $p(Z|X)$ can be further factorized using a probabilistic chain rule and Markov assumption as follows:

$$p(Z|X) = \prod_{t=1}^{T} p(z_t | z_1, \ldots, z_{t-1}, X) \tag{2.32}$$

$$\approx \prod_{t=1}^{T} p(z_t | X) \tag{2.33}$$

The framewise posterior distribution, $p(z_t|X)$ is computed from all inputs, X, and it is directly modeled using BLSTM [35, 45].

$$\boldsymbol{h}_t = BLSTM_t(X) \tag{2.34}$$

$$p(z_t|\, X) = Softmax(LinB(\boldsymbol{h}_t)) \tag{2.35}$$

$BLSTM_t(\cdot)$ takes full input sequence as input and produces hidden vector ($\boldsymbol{h}_t$) at t. Where $Softmax(\cdot)$ represents the softmax activation function. $LinB(\cdot)$ is used to convert the hidden vector, $\boldsymbol{h}_t$, to a $(|\mathcal{U}| + 1)$ dimensional vector with learnable matrix and bias vector parameter.

(b) *CTC letter model:* By applying, Bayes' decision theory probabilistic chain rule, and Markov assumption, $p(C|Z)$ can be written as

$$p(C/Z) = \frac{p(Z/C)p(C)}{p(Z)} \tag{2.36}$$

$$= \prod_{t=1}^{T} p(z_t|\, z_1, \ldots, z_{t-1}, C)\frac{p(C)}{p(Z)} \tag{2.37}$$

$$\approx \prod_{t=1}^{T} p(z_t|z_{t-1}, C)\frac{p(C)}{p(Z)} \tag{2.38}$$

where $p(z_t|z_{t-1}, C)$ represents state transition probability. $p(C)$ represents a letter-based language model, and $p(Z)$ represents the state prior probability. CTC architecture incorporates letter-based language model. CTC architecture can also incorporate a word-based language model by using a letter-to-word finite state transducer during decoding [27]. The CTC has the monotonic alignment property. That is, when $z_{t-1} = c'_m$, then $z_t = c'_l$ where $l \geq m$.

Monotonic alignment property is an important constraint for speech recognition, so ASR sequence-to-sequence mapping should follow the monotonic alignment. This property is also satisfied by HMM/DNN.

(c) *Objective function:* The posterior, $p(C|X)$ is represented as

$$p(C|\, X) \approx \underbrace{\sum z \prod_{t=1}^{T} p(z_t|\, z_{t-1}, C)p(z_t|\, X)}_{\triangleq p_{ctc}(C/X)} \cdot \frac{p(C)}{p(Z)} \tag{2.39}$$

Viterbi method and forward–backward algorithm are dynamic programming algorithms, which are used to efficiently compute the summation over all possible Z. CTC objective function $p_{CTC}(C|X)$ is designed by excluding the $p(C)/p(Z)$ from Eq. (2.23).

The CTC formulation is also the same as HMM/DNN. The minute difference is that Bayes' rule is applied to $p(C|Z)$ instead of $p(W|X)$. It has also three distribution components like HMM/DNN that is, framewise posterior distribution, $p(z_t|X)$, transition probability, $p(z_t|z_{t-1}, C)$, and letter model, $p(C)$. It also uses the Markov assumption. It does not fully utilize the benefit of end-to-end ASR but its character output representation still possesses the end-to-end benefits.

2.4 Convolutional Neural Networks

CNNs are the popular variants of deep learning that are widely adopted in ASR systems. CNNs have many attractive advancements, that is, weight sharing, convolutional filters, and pooling. Therefore CNNs have achieved an impressive performance in ASR. CNNs are composed of multiple convolutional layers. Fig. 2.6 shows the architecture of CNN for speech recognition. LeCun and Bengio [46] describe the three main sublayers of CNN layer, that is, convolution, pooling, and nonlinearity. Each one of the concepts has the potential to boost speech recognition performance.

Locality, weight-sharing, and pooling are the key properties of CNNs that have potential to improve speech recognition performance. Locality offers more robustness against nonwhite noise where some bands are cleaner than the others. The reason is that good features can be computed locally from the cleaner parts of the spectrum and only a smaller number of features are affected by the noise. By this, higher layers of the network get better chance to handle the noise by combining higher level features computed from each frequency band. Locality reduces the number of weights to be learned. Convolutional filters or local filters are imposed on a subset region of the previous layer to capture a specific kind of local patterns known as structural locality. Different phonemes have different energy concentrations in different local bands along the frequency axis. The process of distinguishing different phonemes becomes critical due to these local energy concentrations. These small local energy

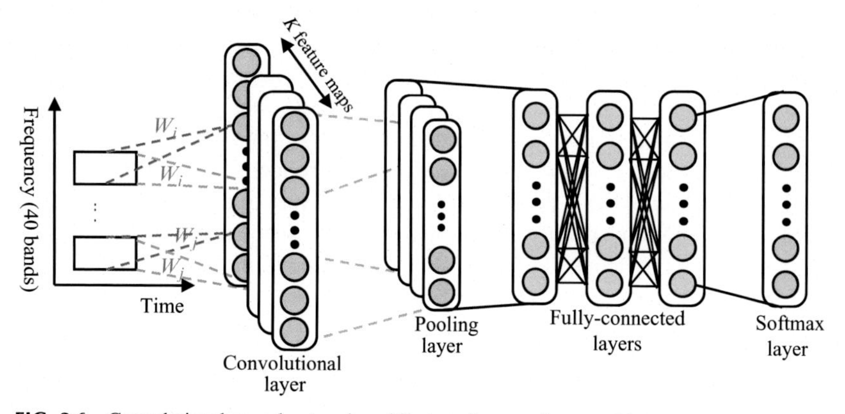

FIG. 2.6 Convolutional neural network architecture for speech recognition.

concentrations also cause the problem of temporal evaluation. These small local energy concentrations of the input are processed by a set of local filters that are applied by a convolutional layer. Local filters are also effective against ambient noises. Weight sharing improves the model robustness by reducing overfitting. It also reduces the number of network weights. The differences in vocal tract lengths among different speaker cause the frequency shifts in speech signals. Weight-sharing carefully handles these frequency shifts, whereas other models like GMMs or DNNs are failed to effectively handle these shifts. In CNN, input is an image of mel-log filter energies visualized as spectrogram. Therefore, CNN ran a small-sized window over the input image; both training and testing time. By this process, the weights of the network learn from the various features of the input data. The properties of the speech signal vary over different frequency bands. Therefore, separate sets of weights for different frequency bands may be more fruitful. This process is known as limited weight sharing. In full weight sharing scheme, convolutional units attached to the same pooling unit share the same convolutional weight. These convolutional units share their weights to compute comparable features.

CNN consists of a special network structure, that is, alternate convolutional and pooling layers. Deep CNNs set a new milestone by achieving approximate human level performance through advanced architectures and optimized training [47]. CNNs use nonlinear function to directly process the low-level data. CNNs are capable of learning high-level features with high complexity and abstraction. Maxout is a widely used nonlinearity that has shown its effectiveness in ASR tasks [48, 49]. Pooling is the heart of CNNs which reduces the dimensionality of a feature map. Pooling causes the minimal differences in the feature extracted because different locations are pooled together. Max pooling is helpful in handling slight shifts in the input pattern along the frequency axis.

Pooling is an important concept that transforms the joint feature representation into the valuable information by keeping the useful information and eliminating insignificant information. Small frequency shifts that arc common in speech signal are efficiently handled using pooling. Pooling also helps in reducing the spectral variance present in the input speech. It maps the input from p adjacent units into the output by applying a special function. After the element-wise nonlinearities, the features are passed through pooling layer. This layer executes the down-sampling on the feature maps coming from the previous layer and produces the new feature maps with a condensed resolution.

The main aim of CNN is to discover local structure in input data. CNN successfully reduces the spectral variations, and it can adequately model the spectral correlations in acoustic features. Earlier the convolution was applied only on frequency axis which made the ASR system more robust against variations generated from different speaking style and speaker [22, 50, 51]. Later, Toth [52] applied convolution along the time axis. Unfortunately, the improvement was negligible. Therefore convolution is generally applied only along the frequency axis because shift-invariance in frequency is more important than shift-invariance in time [51, 53, 54].

CNNs have the capabilities to model temporal correlations with adaptive content information. Therefore they can perform any sequence-to-sequence mapping with great accuracy [55]. CNN layers generate features for succeeding layers that classify them into corresponding classes. The HMMs state emission probability density function is calculated by dividing the state posterior derived from CNNs by the prior probability of the consider state calculated from the training data. The CNN consists of a few sets of convolution and pooling layers. Pooling is generally applied along the frequency domain. Finally, fully connected layers combine inputs from all the positions into a 1-D feature vector. Hence, high-level reasoning is applied on invariant features and then supplied to the softmax, that is, the final layer of the architecture. Finally, the softmax activation layer classifies the overall inputs.

2.4.1 Type of Pooling

CNNs are successfully adopted in many speech recognition tasks and are setting new records. Alternate convolution and pooling layers are followed in CNN architecture, and in last, fully connected layers are used. Pooling is a key-step in CNN-based ASR systems that reduces the dimensionality of the feature maps. Pooling combines a set of values into a smaller number of values. Stride represents the size of pooling, that is, the reduction in dimension of the feature map.

Pooling operators provide a form of spatial transformation invariance, as well as reduce the computational complexity for upper layers by eliminating some connections between convolutional layers. Pooling extracts the output from several regions and combines them. It maps the input from p adjacent units into the output by applying a special function. This layer executes the downsampling on the feature maps coming from the previous layer and produces the new feature maps with a condensed resolution. This layer drastically reduces the spatial dimension of input. It serves two main purposes. First, it reduces the number of parameters or weight by 65%, thus lessening the computational cost. Second, it controls the overfitting. This term refers to when a model is so tuned to the training examples. An ideal pooling method is expected to extract only useful information, and discard irrelevant details. This section covers earlier proposed pooling methods that have been successfully applied in CNNs.

Researchers have proposed many pooling strategies, and in beginning average and max pooling were popular choice [51, 56]. Max and average pooling are deterministic, that is, extract the maximum activation in each pooling region and took the average of all activations in each pooling region, respectively. Average pooling takes all the activations in a pooling region into consideration with equal contributions. This may downplay high activations since many low activations are also included. L_p pooling [57] and stochastic pooling [58] are alternate pooling strategies that address this problem. Bruna et al. [57] claim that the generalization offered by L_p pooling is better than max pooling.

Stochastic pooling selects from a few specific locations (those with strong responses), rather than all possible locations in the region. It forms multinomial distribution [59] by the activations and uses it to select the activation within the pooling region. It uses the stochastic process, hence it is known as stochastic pooling. Mixed pooling [60] is a hybrid of max and average pooling. It can switch between max pooling and average pooling by parameter λ. It takes the advantages of both pooling but both are deterministic in nature, hence it is also deterministic in nature. Multiscale order pooling is applied to improve the invariance of CNNs [61]. Spectral pooling is an advance pooling strategy that crops the features from the input stream [62]. It overcomes the problem of sharp reduction in dimensionality.

2.4.1.1 Max Pooling

The most popular strategy for CNNs is max pooling, which picks only the maximum activation and discards all other units from the pooling region [22]. The function for max pooling is given in Eq. (2.40).

$$s_j = \max_{i \in R_j} a_i \tag{2.40}$$

where R_j is a pooling region and $\{a_1, \ldots, a_{|R_j|}\}$ is a set of activations. Zeiler and Fergus [58] have shown in their experiments that overfitting of the training data is a major problem with max pooling.

2.4.1.2 Average Pooling

Maximum activation of the filter template for each region is selected by the max pooling. However, other activations in the same pooling region are ignored, this should be taken into account. Instead of always extracting maximum activation from each pooling region as max pooling does, the average pooling takes the arithmetic mean of the element in each pooling region. The function for average pooling is given in Eq. (2.41).

$$s_j = \frac{1}{|R_j|} \sum_{i \in R_j} a_i \tag{2.41}$$

The drawback of average pooling is that all the elements in a pooling region are considered, even if many have low magnitude. It weighs down the strong activations because it combines many sparse elements on average.

2.4.1.3 Stochastic Pooling

Inspired by the dropout [63], Zeiler and Fergus [58] proposed the idea of stochastic pooling. In max pooling, the maximum activation is picked from each pooling region. Whereas the areas of high activation are weighed down by areas of low activation in average pooling because all elements in the pooling region are examined, and their average is taken. It is a major problem with average pooling. The issues of max and average pooling are addressed using stochastic pooling. Stochastic pooling applies multinomial distribution [59] to randomly pick the value. It includes the nonmaximal activations of feature maps. In stochastic pooling, first, the probabilities p is computed for each region j by normalizing the activations within the regions, as given in Eq. (2.42).

$$p_i = {}^{a_i}/\sum_{k \in R_j} a_k \tag{2.42}$$

These probabilities create a multinomial distribution that is used to select location l, and corresponding pooled activation a_l is selected based on p. Multinomial distribution selects a location l within the region:

$$s_j = a_l \quad \text{where}\, l \sim P\left(p_1, \ldots, p_{|R_j|}\right)$$

In other words, the activations are selected randomly based on the probabilities calculated by multinomial distribution. Stochastic pooling prohibits overfitting because of the stochastic component. The advantages of max pooling are also found in the stochastic pooling because there's a likelihood of maximum values and a minimal chance of low values. It also utilizes nonmaximal activations. The procedure is shown in Fig. 2.7.

Stochastic pooling represents multinomial distributions of activations within the region, hence the selected element may not be the largest element. It gives high chances to stronger activations and suppresses the weaker activations.

2.4.1.4 L_p Pooling

Sermanet et al. [64] proposed the concept of L_p pooling and claimed that its generalization ability is better than max pooling. In this pooling, a weighted average of inputs is taken in the pooling region. It is shown mathematically below

$$s_j = \left(\frac{1}{|R_j|} \sum_{i \in R_j} a_i^p \right)^{1/p} \tag{2.43}$$

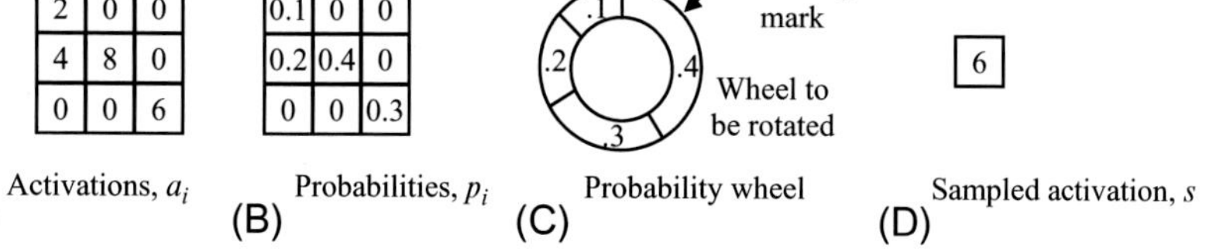

FIG. 2.7 Examples of stochastic pooling: (A) Activations within a given pooling region; (B) probabilities based on activations; (C) probability wheel; (D) sampled activation.

where s_j represents the output of the pooling operator at location j, a_i is the feature value at location i within the pooling region R_j. The value of p varies between 1 and ∞. When $p=1$, L_p operator behaves as average pooling and at $p=\infty$, it leads to max pooling. For L_p pooling, $p>1$ is examined as a trade-off between average and max pooling.

2.4.1.5 Mixed Pooling

Max pooling extracts only the maximum activation, whereas average pooling weighs down the activation by combining the nonmaximal activations. To overcome this problem, Yu et al. [60] proposed a hybrid approach by combining the average pooling and max pooling. This approach is based on Dropout [63] and Dropconnect [65]. Mixed pooling can be represented as

$$s_j = \lambda \max_{i \in R_j} a_j + (1-\lambda)\frac{1}{|R_j|}\sum_{i \in R_j} a_i \tag{2.44}$$

where λ decides the choice of using either max pooling or average pooling. The value of λ is selected randomly in either 0 or 1. When $\lambda=0$, it behaves like average pooling and when $\lambda=1$, it works like max pooling. The value of λ should be recorded during forward-propagation then backpropagation is performed according to the value of λ. Yu et al. [60] showed its superiority over max and average pooling by performing image classification on three datasets.

2.4.1.6 Multiscale Orderless Pooling

Multiscale orderless pooling (MOP) was proposed by Gong et al. [61]. This pooling scheme improves the invariance of CNNs without affecting their discriminative power. MOP processes both whole signal and local patches to extract the deep activation features. The activation features of the whole signal are captured for global spatial layout information and, local patches are captured for more local, fine-grained details of the signal as well as enhancing invariance. Vectors of locally aggregated descriptors encoding [66] is used to aggregate the activation features from local patches. Global spatial activation features and local patch activation features are concatenated using Karhunen-Loeve transformation [67] to obtain the new signal representation.

2.4.1.7 Spectral Pooling

Rippel et al. [62] introduced a new pooling scheme by including the idea of dimensionality reduction by cropping the representation of input in the frequency domain. Let $x \in R^{m \times m}$ be an input feature map and $h \times w$ be the desired dimensions of the output feature map. First, discrete Fourier transform (DFT) [68] is applied on the input feature map, then $h \times w$ size submatrix of frequency representation is cropped from the center. Lastly, inverse DFT is applied on

$h \times w$ submatrix to convert it into spatial domain again. Spectral pooling preserves the most information for the same output dimensionality by tuning the resolution of input precisely to match the desired output dimensionality compared to max pooling. Spectral pooling does not suffer from the sharp reduction in output dimensionality because the stride-based pooling strategy reduces the dimensionality by at least 75% as a function of stride, and it is not the stride-based pooling method. This process is known as low-pass filtering, which exploits the nonuniformity of the spectral density of the data with respect to frequency. It overcomes the problem of sharp reduction in output map dimensionality. The key idea behind spectral pooling is matrix truncation, which reduces its computation cost in CNNs by employing fast Fourier transformation for convolutional kernels [69].

2.4.2 Types of Nonlinear Functions

This section presents the brief idea about different nonlinear functions that are successfully used in various vision tasks with a remarkable result.

2.4.2.1 Sigmoid Neurons

For acoustic modeling, the standard sigmoid is preferred choice. Fixed function shapes and no adaptive parameters are its strengths. The function of the Sigmoid family is thoroughly explored for many ASR tasks.

$$f(\alpha) = \eta \cdot \frac{1}{1 + e^{-\gamma\alpha+\theta}} \tag{2.45}$$

$f(\alpha)$ is the logistic function and θ, γ, and η are known as the learnable parameters. Eq. (2.45) denotes the *p-sigmoid*(η, γ, θ).

In the *p-sigmoid* function, the curve $f(\alpha)$ have different effects of η, γ, and θ. Among the three parameters, the curve $f(\alpha)$ is highly changed by η because it scales linearly. The value of $f(\alpha)$ that is, $|f(\alpha)|$ is always less than or equal to $|\eta|$. η can be any real number. If $\eta < 0$, the hidden unit makes a negative contribution, if $\eta = 0$, the hidden unit is disabled, and if $\eta > 0$, the hidden unit makes a positive contribution that can be seen as a case of linear hidden unit contributions [70, 71]. If $\gamma \longrightarrow 0$, then $f(\alpha)$ is similar to input around 0. The horizontal difference to $f(\alpha)$ is managed using parameter θ. If $\gamma \neq 0$, then $^{\theta}/_{\gamma}$ is the x-value of the mid-point.

2.4.2.2 Maxout Neurons

Maxout neurons came as a good substitute of the sigmoid neurons. The main issue with the use of conventional sigmoid neurons is vanishing gradient problem during stochastic gradient descent (SGD) training. Maxout neurons effectively resolve this issue by producing constant gradient. Cai et al. [72] achieved

1%–5% relative gain using the maxout neurons instead of rectified linear units (ReLU) on the switchboard dataset [73]. Each maxout neuron gets input from several pieces of alternative activations. The maximum value among its piece group is taken as the output of a maxout neuron as given in Eq. (2.46).

$$h_l^{\,i} = \max_{j \in 1,\ldots,k} z_l^{\,ij} \tag{2.46}$$

where $h_l^{\,i}$ represents ith maxout neuron output in the lth layer. k represents the number of activation inputs for the maxout neurons. $z_l^{\,ij}$ represents the jth input activation of the ith neuron in the lth layer as given in Eq. (2.47).

$$z_l = \boldsymbol{W}_l^{\,T} \boldsymbol{h}_{l-1} + \boldsymbol{b}_l \tag{2.47}$$

where $\boldsymbol{W}_l^{\,T}$ represents the transpose of the weight matrix for layer l, $\boldsymbol{h}_{l-1}$ represents previous hidden layers output and $\boldsymbol{b}_l$ is bias vector for layer l and z_l is piece activations (or input activations).

The computation of z_l or $\boldsymbol{h}_l$ does not include any nonlinear transforms like sigmoid or tanh. The process of selection of maximum value, which is the nonlinearity of the maxout neuron, is represented in Eq. (2.46). This process is also known as feature selection process.

During SGD training for the maxout neuron, the gradient is computed as given in Eq. (2.48).

$$\frac{\partial h_l^{\,i}}{\partial z_l^{\,ij}} = \begin{cases} 1 & \text{if } z_l^{\,ij} \geq z_l^{\,is} \forall_s \in 1,\ldots,k \\ 0 & \text{otherwise} \end{cases} \tag{2.48}$$

Eq. (2.48) shows that the value of the gradient is 1 for the neuron's output with the maximum activation and 0 otherwise. The value of gradients is either 0 or 1 during SGD training, so the vanishing gradient problem is resolved easily with maxout neurons. Therefore the deep maxout neural networks are easily optimized as compared to conventional sigmoid neural networks.

2.4.2.3 Rectified Linear Units

Hinton et al. [63] proposed a new way to regularize DNNs by the use of ReLU and dropouts. Regularization is a method to reduce the overfitting in NN by preventing complex coadaptations on the training data. During training, it keeps a neuron active with some probability p, or setting it to zero otherwise. Dahl et al. [74] have shown in their research that a 5% relative reduction has been achieved in WER for cross-entropy trained DNNs using ReLU + dropout on a 50-hour English Broadcast News LVCSR. ReLU is a nonsaturated linear activation function which output zero for negative values and input itself otherwise. Its output is zero for negative values and input itself otherwise. ReLU function is defined as

$$h_l = \max(0, z_l) \tag{2.49}$$

Hessian-free (HF) sequence training works on two key ideas. Initially, error function is calculated by a second-order quadratic at each step then, conjugate Gradient is used to optimize the objective function. It improves the results, but is hard to implement. However, subsequent HF sequence training [75] without dropout erased some of these gains, and performance was the same as provided by a DNN trained with a sigmoid nonlinearity without dropout.

2.4.2.4 Parameterized Rectified Linear Units

ReLU units generate zero gradients whenever the units are not active. Therefore gradient-based optimization will not update their weights. The result of constant zero gradients is a slow down in the training process. To overcome this issue, He et al. [76] proposed parameterized ReLUs (PReLUs) as an advanced version of ReLU. It includes the downside to quicken the learning. It successfully obviates the vanishing gradient problem. In this model, if the input is negative then the output is produced by multiplying input with a constant variable α otherwise the output is the input itself. A PReLU function is defined as follows:

$$h_l = \begin{cases} \alpha z_l & \text{if } z_l < 0 \\ z_l & \text{otherwise} \end{cases} \tag{2.50}$$

2.4.2.5 Dropout

Maxout neurons expertly manage the problem of under-fitting, so they have better optimization performance [77]. However, CNNs using maxout nonlinearity are more prone to overfitting due to their high capability. To address the overfitting problem, regularization methods such as L_p-norm, weight decay, weight tying etc. have been proposed. Hinton et al. [63] proposed a promising regularization technique called "dropout" to efficiently reduce the problem of overfitting. In this method, half of the activations within a layer are stochastically set to zero for each training sample. By doing so, the hidden units cannot coadapt to each other and learn better representation for the inputs. Model averaging is a straightforward generalization method to control the overfitting [78]. Goodfellow et al. [79] showed that dropout is an effective way to control the overfitting for maxout networks because of its better model averaging. Various strategies are used for dropout regularization for the training and testing phase. Specifically, dropout ignores each hidden unit stochastically with probability p during the feed-forward operation in neural network training.

2.5 CNN-Based End-to-End Approach

A novel acoustic model based on CNN is proposed by Palaz et al. [15], which is shown in Fig. 2.8. The CNN-based raw speech phoneme recognition system is composed of three stages. The feature learning stage modeling stage jointly perform the feature extraction and classification task. The third stage is when an

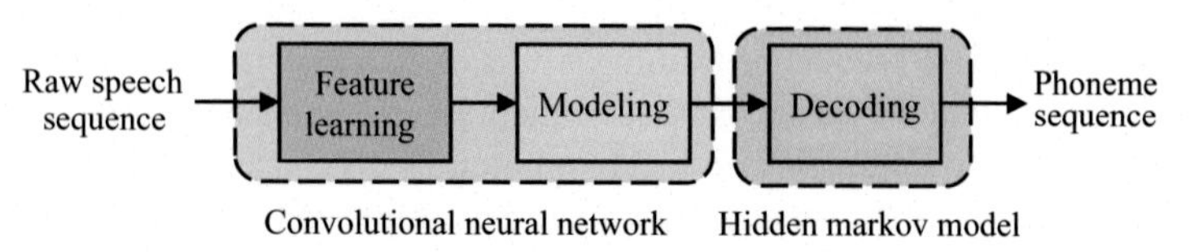

FIG. 2.8 CNN-based raw speech phoneme recognition system.

HMM-based decoder is applied to learn the transition between the different classes. Initially, the raw speech signal is split into features $s_t^c = \{s_{t-c}, \ldots, s_t, \ldots, s_{t+c}\}$ in the context of $2c$ frames with nonoverlapped sliding window w_{in} milliseconds. The first convolutional layer learns the useful features from the raw speech signal and remaining convolutional layers further process these features into useful information. After processing the speech signal, CNN estimates the class conditional probability, for example, $P(i/s_t^c)$ for each class, which is used to calculate emission scaled likelihood $P(s_t^c/i)$. Note: the scaled likelihoods are estimated by dividing the posterior probability by the prior probability of each class, estimated by counting on the training set. Several filter stages are present in the network before the classification stage. A filter stage is a combination of a convolutional layer, a pooling layer, and nonlinearity. The joint training of filter stage and classifier stage is performed using the back-propagation algorithm.

The end-to-end approach employs the following understanding:

1. Speech signals are nonstationary in nature. Therefore they are processed in a short-term manner. Traditional feature extraction methods generally use 20–40 ms sliding window size. Although, in the end-to-end approach, short-term processing of the signal is required. Thus the size of the short-term window is taken as hyperparameter, which is automatically determined during training.
2. Feature extraction is a filter operation because its components like Fourier transform, discrete cosine transform etc. are filtering operations. In traditional systems, filtering is applied on both frequency and time. So, this factor is also considered in building convolutional layer in the end-to-end system. Therefore, the number of filters or kernel, kernel-width, and max-pooling kernel width are taken as hyperparameters that are automatically determined during training. The system automatically checks the different values of hyperparameters to get the optimal values.

The end-to-end model estimates $P(i/s_t^c)$ by processing the speech signal with minimal assumptions or prior knowledge.

2.6 Experiments and Their Results

Texas Instruments, Massachusetts Institute of Technology and SRI International designed (TIMIT) corpus, a popular speech dataset that is widely used for speech research. It is composed of acoustic-phonetic labels consisting of total 6300 utterances from 630 speakers of eight major dialects of American English. There are 6300 utterances that consist of two dialect sentences (SA), 450 phonetically compact sentences (SX) and 1890 phonetically diverse sentences (SI) with the exception of SA sentences which are usually excluded from the tests. The training and test sets do not overlap. It is a well-balanced corpus w.r.t. distribution of phones and triphones. Its full training set contains 4620 utterances, but usually, SI and SX sentences are used. Therefore standard 3696 SI and SX utterances are used for training. A core test set containing 192 utterances from 24 speakers is used for testing, but it is not enough for a reliable conclusion. Hence, the complete test set containing 1344 utterances from 168 speakers is also used for testing purpose in the experiments.

In the first experiment, the filter stage has a number of hyperparameters. w_{in} represents the time span of the input speech signal. kW represents the kernel that is an integral component of the layered architecture. The kernel refers to an operator applied to the input to transform it into features. dW represents the shift of temporal window. kW_{mp} represents the max-pooling kernel width, and dW_{mp} represents the shift of the max-pooling kernel. The value of all hyperparameters is estimated during training based on frame-level classification accuracy on training data. The experiments are conducted for three convolutional layers. The range of hyperparameters after validation is shown in Table 2.2. The best performance for the first convolutional layer is achieved with a kernel width (kW_1) of 50 samples with 310ms of context. These are hyperparameters so their value is achieved during training. Max-pooling is used as the default pooling strategy.

Various pooling strategies that have been successfully applied in computer vision tasks are investigated for speech recognition tasks. A local region of convolutional layer feed the input to the pooling layer that downweigh the input to generate a single output from that region. In the second experiment, pooling

TABLE 2.2 Range of Hyperparameter for TIMIT Dataset During Testing

Hyperparameter	Units	Range
Input window size (w_{in})	ms	100–700
Kernel width of the first ConvNet layer (kW_1)	Samples	10–90
Kernel width of the *n*th ConvNet layer (kW_n)	Samples	1–11
Number of filters per kernel (d_{outt})	Filters	20–100
Max-pooling kernel width (kW_{mp})	Frames	2–6
Number of hidden units in the classifier	Units	200–1500

methods are explored for a CNN-based raw speech recognition model. To check their superiority, all the pooling strategies are evaluated on TIMIT corpus. For rest of the experiments, we used the best performing values for parameters, that is, 150ms of context, 10, 5, and 9 frames kernel width, 10, 1, and 1 frames-shift, 100 filters and 500 hidden units. The results are shown in Table 2.3. The results show that max pooling performs well. However, these gains are not significant for speech recognition tasks as compared to gains achieved in vision tasks [80]. The results are approximately the same, that is, minor difference in recognition rate for all pooling strategies for training and testing set. In the results, MOP and spectral pooling shows a performance drop. The phone error rate (PER) offered by pooling strategies like L_p pooling and stochastic pooling is high as compared to max pooling. Note that MOP and spectral pooling also degrade the performance of ASR systems. The experiments conclude that different pooling strategies are not showing any drastic improvement for speech recognition than it is for other domains such as image classification [81].

In the third experiment, the parameters are the same as the ones used in the second experiment. The performance of different nonlinear functions, for example, sigmoid, maxout, ReLU, and PReLU is evaluated by performing the experiments on fully connected layers with and without using dropout networks on the TIMIT dataset. The experimental results of different nonlinear networks with or without dropout are shown in Table 2.4. Maxout networks converge faster than ReLU, PReLU, and sigmoidal networks. The PER is the same for sigmoid in both cases with or without dropout. During training, the maxout networks show better abilities to fit the training dataset. During testing, the maxout and the PReLU networks have shown almost the same PER, but the sigmoidal and ReLU perform less. Further, the model is evaluated for dropout.

TABLE 2.3 Comparisons of Phone Error Rate (PER) on TIMIT Corpus Development and Test Sets

	PER (%)	
Pooling Strategy	**Training Set**	**Testing**
Max pooling	**19.4**	**21.9**
Average pooling	19.9	22.1
Stochastic pooling	20.0	22.1
L_p pooling	19.7	22.0
Mixed pooling	19.7	22.0
Multiscale orderless pooling	20.1	22.3
Spectral pooling	20.0	22.2

TABLE 2.4 PER as a Function of Nonlinearity

		PER (%)	
Nonlinearity	**Model**	**Training Set**	**Testing Set**
Sigmoid	No-Drop	21.1	22.9
Sigmoid	Dropout	21.0	22.9
Maxout	No-Drop	20.6	22.5
Maxout	**Dropout**	**19.9**	**21.9**
ReLU	No-Drop	21.2	23.0
ReLU	Dropout	20.1	22.2
PReLU	No-Drop	20.6	22.6
PReLU	Dropout	20.0	22.1

TABLE 2.5 Comparison of Existing End-to-End Speech Model in the Context of PER (%)

End-to-End Speech Recognition Model	PER (%)
CNN-based speech recognition system using raw speech as input [17]	33.2
Estimating phoneme class conditional probabilities from raw speech signal using CNNs [41]	32.4
CNNs-based continuous speech recognition using raw speech signal [16]	32.3
End-to-end phoneme sequence recognition using CNNs [15]	27.2
CNN-based direct raw speech model	**21.9**
End-to-end continuous speech recognition using attention-based recurrent NN: first results [28]	18.57
Towards end-to-end speech recognition with deep CNNs [49]	18.2
Attention-based models for speech recognition [7]	17.6
Segmental RNNs for end-to-end speech recognition [82]	17.3

The same dropout rate is applied for each layer. Dropout is varied with the step size 0.05. The experiments confirmed that the dropout probability $p=0.5$ is reasonable.

Table 2.5 shows the comparison of existing end-to-end speech recognition model in the context of PER. The evaluated model is not the best model, but it is

TABLE 2.6 Comparison of Existing Techniques with CNN-Based Direct Raw Speech Model in the Context of PER (%)

Methods	PER (%)
GMM/HMM-based ASR system [83]	34
CNN-based direct raw speech model	**21.9**
Attention-based models for speech recognition [7]	17.6
Segmental RNNs for end-to-end speech recognition [82]	17.3
Combining time and frequency domain convolution in CNN-based phone recognition [52]	16.7
Phone recognition with hierarchical convolutional deep maxout networks [84]	16.5

satisfactory. The result of the experiment conducted on the TIMIT dataset for this model is compared with already existing techniques, and they are shown in Table 2.6. The main advantages of this model is that it offers close performance by directly using the raw speech. It also increases the generalization capability of the classifiers.

2.7 Conclusion

The CNN-based direct raw speech recognition model directly learns the relevant representation from the speech signal in a data-driven method and calculates the conditional probability for each phoneme class. In this, CNN as an acoustic model, consists of a feature stage and classifier stage. Both the stages are trained jointly. Raw speech is supplied as input to the first convolutional layer, and several convolutional layers further process it. Classifiers like ANN, CRF, MLP or fully connected layers, calculate the conditional probabilities for each phoneme class. Afterwards, decoding is performed using HMM. This model shows the better performance as shown by MFCC-based conventional model.

References

[1] L.R. Rabiner, B.-H. Juang, Fundamentals of Speech Recognition, vol. 14, Prentice Hall, Englewood Cliffs, NJ, 1993.

[2] S.B. Davis, P. Mermelstein, Comparison of parametric representations for monosyllabic word recognition in continuously spoken sentences, in: Readings in Speech Recognition, Elsevier, 1990, pp. 65–74.

[3] H. Hermansky, Perceptual linear predictive (PLP) analysis of speech, J. Acoust. Soc. Am. 87 (1990) 1738–1752.
[4] A. Graves, A.-r. Mohamed, G. Hinton, Speech recognition with deep recurrent neural networks, in: Presented at the 2013 IEEE International Conference on Acoustics, Speech and Signal Processing, 2013.
[5] A. Graves, N. Jaitly, Towards end-to-end speech recognition with recurrent neural networks, in: International Conference on Machine Learning, 2014, pp. 1764–1772.
[6] W. Chan, N. Jaitly, Q. Le, O. Vinyals, Listen, attend and spell: a neural network for large vocabulary conversational speech recognition, in: 2016 IEEE International Conference on Acoustics, Speech and Signal Processing (ICASSP), 2016, pp. 4960–4964.
[7] J.K. Chorowski, D. Bahdanau, D. Serdyuk, K. Cho, Y. Bengio, Attention-based models for speech recognition, in: Advances in Neural Information Processing Systems, 2015, pp. 577–585.
[8] H. Soltau, H.-K. Kuo, L. Mangu, G. Saon, T. Beran, Neural network acoustic models for the DARPA RATS program, in: INTERSPEECH, 2013, pp. 3092–3096.
[9] L. Lu, X. Zhang, S. Renais, On training the recurrent neural network encoder-decoder for large vocabulary end-to-end speech recognition, in: 2016 IEEE International Conference on Acoustics, Speech and Signal Processing (ICASSP), 2016, pp. 5060–5064.
[10] J. Chorowski, N. Jaitly, Towards better decoding and language model integration in sequence to sequence models, arXiv: 1612.02695(2016).
[11] H. Soltau, H. Liao, H. Sak, Neural speech recognizer: acoustic-to-word LSTM model for large vocabulary speech recognition, arXiv: 1610.09975(2016).
[12] A. Graves, S. Fernández, F. Gomez, J. Schmidhuber, Connectionist temporal classification: labelling unsegmented sequence data with recurrent neural networks, in: Proceedings of the 23rd International Conference on Machine Learning, 2006, pp. 369–376.
[13] R. Prabhavalkar, K. Rao, T.N. Sainath, B. Li, L. Johnson, N. Jaitly, A comparison of sequence-to-sequence models for speech recognition, in: Proceedings of INTERSPEECH, 2017, pp. 939–943.
[14] B. Li, T.N. Sainath, K.C. Sim, M. Bacchiani, E. Weinstein, P. Nguyen, et al., Multi-dialect speech recognition with a single sequence-to-sequence model, arXiv: 1712.01541(2017).
[15] D. Palaz, R. Collobert, M.M. Doss, End-to-end phoneme sequence recognition using convolutional neural networks, arXiv: 1312.2137(2013).
[16] D. Palaz, M. Magimai-Doss, R. Collobert, Convolutional neural networks-based continuous speech recognition using raw speech signal, in: 2015 IEEE International Conference on Acoustics, Speech and Signal Processing (ICASSP), 2015, pp. 4295–4299.
[17] D. Palaz, M. Magimai-Doss, R. Collobert, Analysis of cnn-based speech recognition system using raw speech as input, in: Proceedings of INTERSPEECH, Idiap, 2015.
[18] D. Povey, Discriminative Training for Large Vocabulary Speech Recognition, University of Cambridge, 2005.
[19] G. Hinton, L. Deng, D. Yu, G.E. Dahl, A.-r. Mohamed, N. Jaitly, et al., Deep neural networks for acoustic modeling in speech recognition: the shared views of four research groups, IEEE Signal Process. Mag. 29 (2012) 82–97.
[20] G.E. Dahl, D. Yu, L. Deng, A. Acero, Context-dependent pre-trained deep neural networks for large-vocabulary speech recognition, IEEE Trans. Audio Speech Lang. Process. 20 (2012) 30–42.
[21] F. Seide, G. Li, D. Yu, Conversational speech transcription using context-dependent deep neural networks, in: Twelfth Annual Conference of the International Speech Communication Association, 2011.

[22] O. Abdel-Hamid, A.-r. Mohamed, H. Jiang, G. Penn, Applying convolutional neural networks concepts to hybrid NN-HMM model for speech recognition, in: Presented at the 2012 IEEE International Conference on Acoustics, Speech and Signal Processing (ICASSP), 2012.

[23] A. Senior, G. Heigold, M. Bacchiani, H. Liao, GMM-free DNN acoustic model training, in: 2014 IEEE International Conference on Acoustics, Speech and Signal Processing (ICASSP), 2014, pp. 5602–5606.

[24] M. Bacchiani, A. Senior, G. Heigold, Asynchronous, online, GMM-free training of a context dependent acoustic model for speech recognition, in: Fifteenth Annual Conference of the International Speech Communication Association, 2014.

[25] M. Gales, S. Young, The application of hidden Markov models in speech recognition, Found. Trends Signal Process. 1 (2008) 195–304.

[26] A. Hannun, C. Case, J. Casper, B. Catanzaro, G. Diamos, E. Elsen, et al., Deepspeech: scaling up end-to-end speech recognition, arXiv: 1412.5567(2014).

[27] Y. Miao, M. Gowayyed, F. Metze, EESEN: End-to-end speech recognition using deep RNN models and WFST-based decoding, in: 2015 IEEE Workshop on Automatic Speech Recognition and Understanding (ASRU), 2015, pp. 167–174.

[28] J. Chorowski, D. Bahdanau, K. Cho, Y. Bengio, End-to-end continuous speech recognition using attention-based recurrent NN: First results, arXiv: 1412.1602(2014).

[29] D. Bahdanau, J. Chorowski, D. Serdyuk, P. Brakel, Y. Bengio, End-to-end attention-based large vocabulary speech recognition, in: 2016 IEEE International Conference on Acoustics, Speech and Signal Processing (ICASSP), 2016, pp. 4945–4949.

[30] W. Chan, I. Lane, On online attention-based speech recognition and joint mandarin character-pinyin training, in: INTERSPEECH, 2016, pp. 3404–3408.

[31] P. Golik, Z. Tüske, R. Schlüter, H. Ney, Convolutional neural networks for acoustic modeling of raw time signal in LVCSR, in: Sixteenth Annual Conference of the International Speech Communication Association, 2015.

[32] A. Graves, Generating sequences with recurrent neural networks, arXiv: 1308.0850(2013).

[33] D. Bahdanau, K. Cho, Y. Bengio, Neural machine translation by jointly learning to align and translate, arXiv: 1409.0473(2014).

[34] V. Mnih, N. Heess, A. Graves, Recurrent models of visual attention, in: Advances in Neural Information Processing Systems, 2014, pp. 2204–2212.

[35] A. Graves, N. Jaitly, A.-r. Mohamed, Hybrid speech recognition with deep bidirectional LSTM, in: 2013 IEEE Workshop on Automatic Speech Recognition and Understanding (ASRU), 2013, pp. 273–278.

[36] A.Y. Hannun, A.L. Maas, D. Jurafsky, A.Y. Ng, First-pass large vocabulary continuous speech recognition using bi-directional recurrent DNNs, arXiv: 1408.2873(2014).

[37] N. Jaitly, G. Hinton, Learning a better representation of speech soundwaves using restricted Boltzmann machines, in: 2011 IEEE International Conference on Acoustics, Speech and Signal Processing (ICASSP), 2011, pp. 5884–5887.

[38] Z. Tüske, P. Golik, R. Schlüter, H. Ney, Acoustic modeling with deep neural networks using raw time signal for LVCSR, in: Fifteenth Annual Conference of the International Speech Communication Association, 2014.

[39] T.N. Sainath, O. Vinyals, A. Senior, H. Sak, Convolutional, long short-term memory, fully connected deep neural networks, in: 2015 IEEE International Conference on Acoustics, Speech and Signal Processing (ICASSP), 2015, pp. 4580–4584.

[40] Y. LeCun, Generalization and network design strategies, in: Connectionism in Perspective, 1989, pp. 143–155.

[41] D. Palaz, R. Collobert, M.M. Doss, Estimating phoneme class conditional probabilities from raw speech signal using convolutional neural networks, arXiv: 1304.1018(2013).
[42] L.R. Rabiner, B.-H. Juang, Speech recognition: statistical methods, in: Encyclopedia of Linguistics, 2006, pp. 1–18.
[43] H.A. Bourlard, N. Morgan, Connectionist Speech Recognition: A Hybrid Approach, vol. 247, Springer Science & Business Media, 2012.
[44] T. Mikolov, M. Karafiát, L. Burget, J. Černocký, S. Khudanpur, Recurrent neural network based language model, in: Eleventh Annual Conference of the International Speech Communication Association, 2010.
[45] S. Hochreiter, J. Schmidhuber, Long short-term memory, Neural Comput. 9 (1997) 1735–1780.
[46] Y. LeCun, Y. Bengio, Convolutional networks for images, speech, and time series, in: The Handbook of Brain Theory and Neural Networks, vol. 3361, 1995, p. 1995.
[47] A. Krizhevsky, I. Sutskever, G.E. Hinton, Imagenet classification with deep convolutional neural networks, in: Advances in Neural Information Processing Systems, 2012, pp. 1097–1105.
[48] X. Zhang, J. Trmal, D. Povey, S. Khudanpur, Improving deep neural network acoustic models using generalized maxout networks, in: 2014 IEEE International Conference on Acoustics, Speech and Signal Processing (ICASSP), 2014, pp. 215–219.
[49] Y. Zhang, M. Pezeshki, P. Brakel, S. Zhang, C.L.Y. Bengio, A. Courville, Towards end-to-end speech recognition with deep convolutional neural networks, arXiv: 1701.02720(2017).
[50] O. Abdel-Hamid, L. Deng, D. Yu, Exploring convolutional neural network structures and optimization techniques for speech recognition, in: INTERSPEECH, 2013, pp. 3366–3370.
[51] T.N. Sainath, A.-r. Mohamed, B. Kingsbury, B. Ramabhadran, Deep convolutional neural networks for LVCSR, in: Presented at the 2013 IEEE International Conference on Acoustics, Speech and Signal Processing, 2013.
[52] L. Toth, Combining time- and frequency-domain convolution in convolutional neural network-based phone recognition, in: Presented at the 2014 IEEE International Conference on Acoustics, Speech and Signal Processing (ICASSP), 2014.
[53] O. Abdel-Hamid, A.R. Mohamed, H. Jiang, L. Deng, G. Penn, D. Yu, Convolutional neural networks for speech recognition, IEEE Trans. Audio Speech Lang. Process. 22 (Oct 2014) 1533–1545.
[54] T.N. Sainath, B. Kingsbury, A.-r. Mohamed, G.E. Dahl, G. Saon, H. Soltau, et al., Improvements to deep convolutional neural networks for LVCSR, in: Presented at the 2013 IEEE Workshop on Automatic Speech Recognition and Understanding, 2013.
[55] J. Gehring, M. Auli, D. Grangier, D. Yarats, Y.N. Dauphin, Convolutional sequence to sequence learning, arXiv: 1705.03122(2017).
[56] Y. LeCun, F.J. Huang, L. Bottou, Learning methods for generic object recognition with invariance to pose and lighting, in: Proceedings of the 2004 IEEE Computer Society Conference on Computer Vision and Pattern Recognition, 2004. CVPR 2004, 2004, pp. II–104.
[57] J. Bruna, A. Szlam, Y. LeCun, Signal recovery from pooling representations, arXiv: 1311.4025 (2013).
[58] M.D. Zeiler, R. Fergus, Stochastic pooling for regularization of deep convolutional neural networks, arXiv: 1301.3557(2013).
[59] E.W. Weisstein, Multinomial series, Available athttp://mathworld.wolfram.com/MultinomialSeries.html.
[60] D. Yu, H. Wang, P. Chen, Z. Wei, Mixed pooling for convolutional neural networks, in: Rough Sets and Knowledge Technology, Springer, 2014, pp. 364–375.

[61] Y. Gong, L. Wang, R. Guo, S. Lazebnik, Multi-scale orderless pooling of deep convolutional activation features, in: European Conference on Computer Vision, 2014, pp. 392–407.
[62] O. Rippel, J. Snoek, R.P. Adams, Spectral representations for convolutional neural networks, in: Advances in Neural Information Processing Systems, 2015, pp. 2449–2457.
[63] G.E. Hinton, N. Srivastava, A. Krizhevsky, I. Sutskever, R.R. Salakhutdinov, Improving neural networks by preventing co-adaptation of feature detectors, arXiv: 1207.0580(2012).
[64] P. Sermanet, S. Chintala, Y. LeCun, Convolutional neural networks applied to house numbers digit classification, in: 2012 21st International Conference on Pattern Recognition (ICPR), 2012, pp. 3288–3291.
[65] L. Wan, M. Zeiler, S. Zhang, Y. Le Cun, R. Fergus, Regularization of neural networks using dropconnect, in: International Conference on Machine Learning, 2013, pp. 1058–1066.
[66] H. Jegou, F. Perronnin, M. Douze, J. Sanchez, P. Perez, C. Schmid, Aggregating local image descriptors into compact codes, IEEE Trans. Pattern Anal. Mach. Intell. 34 (Sep 2012) 1704–1716.
[67] R. Dony, Karhunen-loeve transformThe Transform and Data Compression Handbook, vol. 1, (2001) pp. 1–34.
[68] P. Duhamel, B. Piron, J.M. Etcheto, On computing the inverse DFT, IEEE Trans. Acoust. Speech Signal Process. 36 (1988) 285–286.
[69] M. Mathieu, M. Henaff, Y. LeCun, Fast training of convolutional networks through ffts, arXiv: 1312.5851(2013).
[70] E. Variani, T. Schaaf, VTLN in the MFCC domain: Band-limited versus local interpolation, in: Twelfth Annual Conference of the International Speech Communication Association, 2011.
[71] G. Saon, H. Soltau, A. Emami, M. Picheny, Unfolded recurrent neural networks for speech recognition, in: Fifteenth Annual Conference of the International Speech Communication Association, 2014.
[72] M. Cai, Y. Shi, J. Liu, Deep maxout neural networks for speech recognition, in: 2013 IEEE Workshop on Automatic Speech Recognition and Understanding (ASRU), 2013, pp. 291–296.
[73] J.J. Godfrey, E.C. Holliman, J. McDaniel, SWITCHBOARD: telephone speech corpus for research and development, in: 1992 IEEE International Conference on Acoustics, Speech, and Signal Processing, ICASSP-92, 1992, pp. 517–520.
[74] G.E. Dahl, T.N. Sainath, G.E. Hinton, Improving deep neural networks for LVCSR using rectified linear units and dropout, in: 2013 IEEE International Conference on Acoustics, Speech and Signal Processing (ICASSP), 2013, pp. 8609–8613.
[75] B. Kingsbury, T.N. Sainath, H. Soltau, Scalable minimum Bayes risk training of deep neural network acoustic models using distributed hessian-free optimization, in: Thirteenth Annual Conference of the International Speech Communication Association, 2012.
[76] K. He, X. Zhang, S. Ren, J. Sun, Delving deep into rectifiers: Surpassing human-level performance on imagenet classification, in: Proceedings of the IEEE International Conference on Computer Vision, 2015, pp. 1026–1034.
[77] L. Tóth, Convolutional deep maxout networks for phone recognition, in: Fifteenth Annual Conference of the International Speech Communication Association, 2014.
[78] J. Li, X. Wang, B. Xu, Understanding the dropout strategy and analyzing its effectiveness on LVCSR, in: 2013 IEEE International Conference on Acoustics, Speech and Signal Processing (ICASSP), 2013, pp. 7614–7618.
[79] I.J. Goodfellow, D. Warde-Farley, M. Mirza, A. Courville, Y. Bengio, Maxout networks, arXiv: 1302.4389(2013).
[80] J.S. Ren, L. Xu, On vectorization of deep convolutional neural networks for vision tasks, in: AAAI, 2015, pp. 1840–1846.

[81] L. Liu, C. Shen, A. van den Hengel, Cross-convolutional-layer pooling for image recognition, IEEE Trans. Pattern Anal. Mach. Intell. 39 (2017) 2305–2313.

[82] L. Lu, L. Kong, C. Dyer, N.A. Smith, S. Renals, Segmental recurrent neural networks for end-to-end speech recognition, in: Presented at the Interspeech 2016, 2016.

[83] F. Fauziya, G. Nijhawan, A comparative study of phoneme recognition using GMM-HMM and ANN based acoustic modeling, Int. J. Comput. Appl. 98 (2014).

[84] L. Toth, Phone recognition with hierarchical convolutional deep maxout networks, Eurasip J. Audio Speech Music Process. 2015 (Sep 4 2015).

Chapter 3

A Real-Time DSP-Based System for Voice Activity Detection and Background Noise Reduction

Charu Singh*, Maarten Venter*, Rajesh Kumar Muthu† and David Brown*

*Sat-Com (PTY) Ltd., Windhoek, Namibia, †Vellore Institute of Technology, Vellore, India

3.1 Introduction

These days, most of the young generation are using headphones or earphones to listen to music in public places such as, malls, airports, railway stations, etc. These devices cause sensory and cognitive distractions and isolate the wearer from the external environment. Environment noise plays an important role in the design of speech processing algorithms. For example, when train departures are announced in a noisy railway station, people wearing headphones or earphones may miss this announcement due to the environment noise. In military applications, the important announcement may be missed due to the noisy environment during combat.

Digital signal processors (DSP) can be well integrated into several areas such as engineering, mathematics, and science. The advancement and the development of methods in microelectronics led to the analysis of continuous-time signals, which was facilitated by efficient mathematics. Thus for the processing of signals, which are periodic in nature, the Fourier transform is an invaluable tool. It's also important for work with aperiodic signals. The number of multiplications reduced drastically with the use of DFT (discrete Fourier transform) and fast Fourier transform (FFT). It is known that in a noisy military environment, communication is affected by new and unknown noises. Therefore the single noise compensation algorithm cannot be expected to work efficiently and effectively and minimize all the noise from various sources. To perform speech noise detection and reduction in noisy environmental conditions, the speech signal must be studied and analyzed from many different perspectives relative to the military application.

Intelligent Speech Signal Processing. https://doi.org/10.1016/B978-0-12-818130-0.00003-9

There are currently a large number of analog or continuous signals that vary between minimum and maximum values such as speech, atmospheric pressure, relative humidity, temperature, and light. The DSP is an integrated circuit that comprises a digital processor and a set of complementary resources. It is capable to digitally handle analog signals from the real world like speeches, sounds, and images. At the beginning of the 80s, of the last century, several models of DSPs were already commercialized by Texas Instruments, Nec, and Intel. Those companies are making more powerful models to be coupled to the complex applications.

Price et al. [1] propose an architecture that uses deep neural networks (DNN) for automatic speech recognition (ASR) and voice activity detection (VAD) with improved accuracy, programmability, and scalability. The architecture is designed to minimize off-chip memory bandwidth, which is the main driver of system power consumption. Xu et al. [2] developed a VAD algorithm that is robust to background noise. The algorithm calculates permutation entropy and determines the presence or absence of speech, as well as distinguishing between voiced and unvoiced parts of speech. Experiments done under several noise cases demonstrated that the proposed method can obtain conspicuous improvements on the aspect of false alarm rates, while maintaining comparable speech detection rates when compared to the reference method.

According to Dey and Ashour [3], the speech signals broadcast can be measured as functions of time variables with conserved signal information through the propagation. In order to renovate the signal, it is required to temporally sample the band-limited signal at the specific spatially, and to sample the band-limited signal at the specific time instant. Dey et al. [4, pp., 21–31] discussed the fundamentals of the acoustics with a detailed explanation of several body acoustic sounds sources. Dey et al. [5, pp., 43–47] introduce some examples of acoustic sensors in different biomedical applications.

The problem with acoustic source localization and tracking (ASLT) in noisy environments, while using many microphones, presents a number of challenging difficulties [6]. The presence of silence gaps in speech can easily misguide. This temporally discontinuous nature of speech signals is one of the main challenges when considering real-world situations involving human speakers. Further, this study presented a real-time implementation of the ASLT mechanism for low-power embedded VAD systems.

In order to minimize the possible hazard of natural disasters of the geophysical environment, real-time event detection is required for warnings using early warning monitoring systems. The study implemented a real-time long period and a volcano-tectonic event detector based on VAD systems [7]. This study proposed a miniature DSP that is based on the real-time VAD algorithm for earphones or headphones [8]. This system allows music listener by headphone or microphone to hear external speech signals like public announcements without removing their music player devices.

VADs are often used to identify the part of the speech signal activity and noisy speech signals in speech activity. To perform VAD in real time, a smartphone app using convolutional neural network was developed with low audio latency [9]. An improved VAD algorithm was proposed using frequency analysis including spectral entropy, spectral flatness, and harmonic spectral peaks [10]. This method was tested in three noise environments and compared with various existing works.

However, speech communication in the presence of background noise is the major issue for military personnel who face challenging high-noise conditions during training exercises, inside or near military vehicles in combat zones. This chapter proposed an improved DSP-based system for VAD and background noise reduction for real-time applications; this method can be used in military applications.

3.2 Microchip dsPIC33 Digital Signal Controller

Microchip Technology Inc. introduced Microchip dsPIC33 DSCs (digital signal controllers). This DSC can be programmed easily for specific military and other communications applications without program complexity of DSP.

It provides a cost-effective solution for modern applications like two-way radios, answering machines, hands-free kits, optical networking, power-line modems, audio playback applications, portable medical equipment, security systems, fire alarms, and speech.

The features of Microchip dsPIC33 are shown in Fig. 3.1. dsPIC33 product families are ideal for a wide variety of 16-bit MicroController Unit (MCU) embedded applications. The digital signal processor dsPIC33 device family employs a powerful 16-bit architecture and is used as a control controller [11]. The dsPIC33 series is the second generation of the dsPIC30F series. It

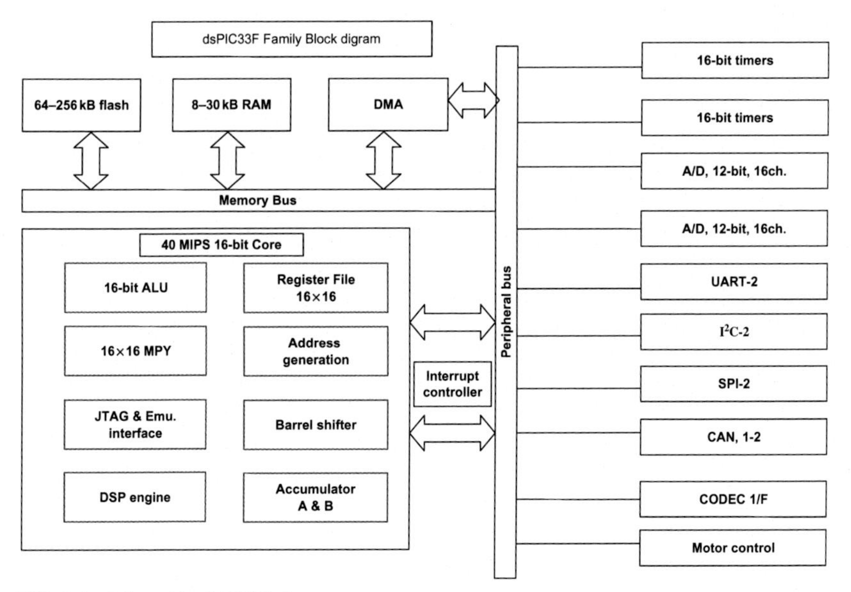

FIG. 3.1 Microchip dsPIC33 features.

is an improved version of it and it is fully compatible with the dsPIC30F version. The differences are mainly in the power supply, FLASH and RAM expansion, the addition of DMA (Direct Memory Access) controller, which provides direct access to data, etc. Microchip also extended the functionality of supported communication protocols and other peripherals. One of the biggest attractions is the dsPIC33 series does not have an integrated EEPROM memory, but the program memory can be used as a data flash. Compared to the dsPIC30F at a lower voltage, it achieves more power, 10 MIPS units at operating temperature (−40° C to 85°C) and the processor voltage (3.0–3.6 V) [11]. Further study can be extended for the project set up the firmware for the dsPIC33 series microprocessor to develop the DSP-based VAD and background noise reduction method.

The speed of the Microchip dsPIC33 is 40 MIPS and it includes 64 Kbps to 256 Kbytes on-chip self-programming Flash, 8–30 Kbytes of RAM, analog-to-digital converter (ADC) capable of converting analog input signals into 12-bit digital words at rates 1.1 MSPS (mega sample per seconds), high performance ADCs, and an 8-channel nonintrusive DMA [12].

The programming of DSPs is too difficult for typical interrupt-driven embedded control applications and most of the professionals are not familiar with DSPs programming. Therefore there is a need for tools to address the concern. The dsPIC33 addresses these issues by providing Microchip's MPLAB Integrated Development Environment (IDE), and C compiler.

Applications for the Microchip dsPIC33 motor-control family include sewing machines, washing machines, LED lighting arrays, online UPS, access control, electronically assisted power steering, precision manufacturing equipment, environmental control, inverters, electric vehicles, absolute encoders, and resolvers.

3.2.1 VAD and Noise Suppression Algorithm

The incoming speech signal is susceptible to ambient noise for microphone-based applications. The Microchip dsPIC33 Noise Suppression Library is useful for speech communication and applications like speakerphone, radios, intercom, and military communications systems to reduce unwanted noise [12]. The DSP instruction set was written in assembly language for the dsPIC33 Noise Suppression Library to make efficient use of the dsPIC33 DSC device architecture, including advanced addressing modes. The dsPIC33 Noise Suppression Library function can be called by user-defined application programmer's interface (API).

Fig. 3.2 shows the VAD system. The dsPIC33 Noise Suppression Library function uses 16-bit speech data sampled at 8 kHz and this function eliminates noise from 10 ms block. To analyze the frequency component of the speech signal, this library function performed FFT on each 10 ms block of data. To determine the nature of speech and whether it is a signal segment noise, a VAD algorithm is used after FFT.

When a noise segment is detected by the VAD, the dsPIC33 noise suppression algorithm updates and maintains a profile of the noise. According to the

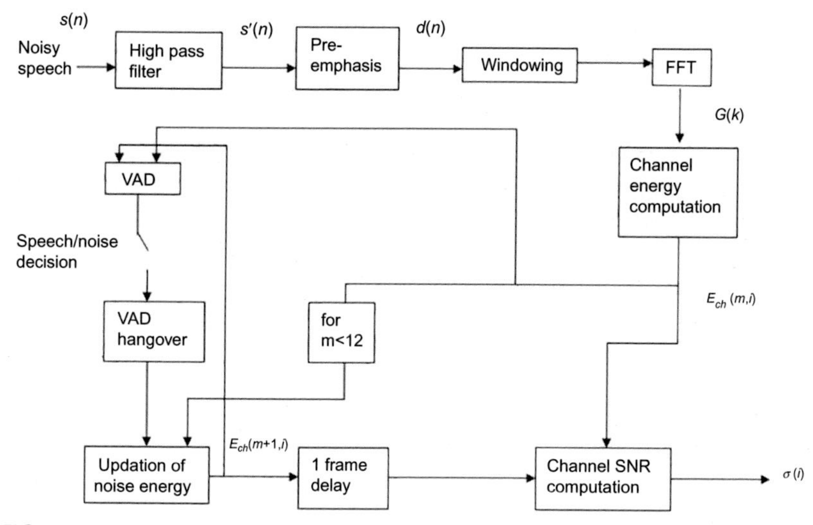

FIG. 3.2 Voice activity detection system.

amount of noise contained in the frequency band, the input signal is scaled for that frequency band.

According to the proportion of noise contained in the frequency band, the input signal is scaled for that frequency band as shown in Fig. 3.3.

As shown in Figs 3.2 and 3.3, the dsPIC33 VAD and noise suppression algorithm can be generally divided into a high pass filter (HPF), FFT, band energy computation, band signal-to-noise ratio (SNR) computation, scale factor computation, VAD, and noise band energy computation, scaling of frequency bands, and time domain conversion functions.

A typical noise suppression algorithm includes an input signal, which is generally a noise-corrupted speech signal, and an output signal that includes the input speech signal with its noise component suppressed.

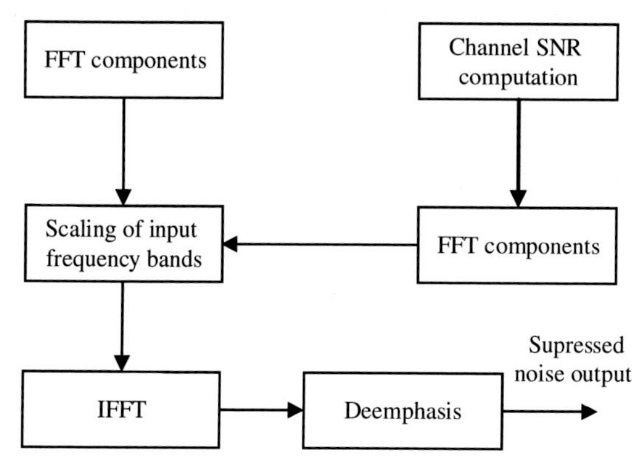

FIG. 3.3 Noise reduction algorithm.

3.3 High Pass Filter

The real-time analog signals are passed through a digital-analog converter so that the digitized signals are introduced to the processors. After the signal is processed, the signal is converted back to its analog form with the ADC.

In this research project study, a HPF is used in the developed mechanism. In digital processing circuits, the HPFs are formed by RC circuit and output is taken from across the register R. The output voltage of all the frequencies below the cut-off frequency is attenuated and the magnitude of the output voltage is constant for above the cut-off frequency.

An HPF is used to remove DC components and other low-frequency components. In order to filter low-frequency components, an HPF with cut-off frequency 80 Hz is applied.

The HPF transfer function is calculated by the following equation:

$$H(z) = 0.5\frac{0.92727435 - 1.8544941z^{-1} + 0.92727435z^{-2}}{1 - 1.9059465z^{-1} + 0.9114024z^{-2}} \tag{3.1}$$

3.4 Fast Fourier Transform

The input samples of the speech signal are preemphasized using the following equation:

$$d(n) = s'(n) + \xi s'(n-1) \tag{3.2}$$

where $\xi = -0.8$.

Preemphasis of the input audio signals is used to improve the SNR by increasing the magnitude of higher frequencies with respect to the magnitude of lower frequencies (within the frequency band). Thus it minimizes the adverse effects of signal distortion, attenuation, or saturation.

The trapezoidal windowing technique is applied to the preemphasized signal. The signals are buffered for 10 ms and (a zero padding of 24 zeroes) overlapped with previous frames of 3 ms. A 64-point complex FFT is performed on the real-time input signal (128-point buffer).

3.5 Channel Energy Computation

Frequency components after FFT are divided into 16 nonuniform channels also known as bands. Energy in each channel is computed by using leaky integration as shown in the equation:

$$E_{ch}(m,i) = \max\left\{ E_{min}, \alpha_{ch}E_{ch}(m-1),i) + (1-\alpha_{ch})\frac{\sum_{k=f_L(i)}^{k=f_H(i)} |G(k)|^2}{f_H(i) - f_L(i) + 1} \right. \tag{3.3}$$

For the first frame the value of α_{ch} is zero and 0.55 for all other frames. The value of E_{min} is 0.0625, and FFT of the input signal is represented by G_K and also current R_{IN} frame represented by m.

3.6 Channel SNR Computation

For each of the frequency bands, SNR can be computed by using

(i) The energy of the channel computed in channel energy computation stage.
(ii) The energy of noise computed in noise energy computation stage (updated only when frames are detected as noise only frames during VAD).

The scale factor for the channel (i) is given by:

$$\sigma(i) = \max\left\{0, \mathrm{round}\left(10\log\left(\frac{E_{ch}(m,i)}{E_n(m,i)}\right)\right\}\right. \tag{3.4}$$

where noise energy estimation for the frame (m) is $E_n(m, i)$.

3.7 VAD Decision

The comparison is made between the channel energies of the current frame and noise channel energies. The output of the VAD decision is the speech flag, which conveys to us whether the current frame is a noisy frame or a speech frame; this decision is based on the number of hangover frames applied.

3.8 VAD Hangover

VAD is sometimes unable to detect the end part of some words or also some silent part that occurs in the middle. This problem can be solved by hysteresis; we add a VAD hangover of some frames. Low-power ending of speech bursts is difficult to detect, therefore hangover makes it easy to detect it. A hangover from 0 to 100 frames can be applied.

3.9 Computation of Scaling Factor

The signal is multiplied by a scale factor in each frequency band. The scaling factor is obtained by subtracting a constant term of 15 dB from the SNR obtained in the channel SNR computation stage. Each scaling factor is converted from a logarithmic scale to a linear scale. It is computed by using the formula:

$$\gamma_{\mathrm{dB}}^{i}(m) = \min(\sigma(i) - 15, 0) \tag{3.5}$$

Computed values can be converted in the linear scale as

$$\gamma_{ch}^{i}(m) = 10^{\gamma_{\mathrm{dB}}^{i}(m)/20} \tag{3.6}$$

3.10 Scaling of Frequency Channels

The components obtained from the FFT output within each frequency channel are multiplied by the factor computed in scaling factor computation stage, known as a frequency domain filtering operation. Within a channel i, for each frequency component k, the scaled output $H(k)$ is given by

$$H(k) = \gamma_{ch}(i) \cdot G(k); f_L(i) \le k \le f_H(i), 0 \le 1 < N_c \tag{3.7}$$

$$= G(k); 0 \le f_L(0), f_H(N_c - 1) < k \le \frac{m}{2} \tag{3.8}$$

3.11 Inverse Fourier Transform

The noise-suppressed output signal can be obtained by converting back into time domain from frequency domain using IFFT. The output signals are de-emphasized using the equation:

$$h'(\mathrm{n}) = h(m, n) + h(m-1, n+L); 0 \le n < M - L \tag{3.9}$$

$$h'(n) = h(m, n); M - L \le n < L \tag{3.10}$$

$$s''(n) = h'(n) + \varepsilon s''(n-1); 0 \le n < L \tag{3.11}$$

where the value of the ε is 0.8 and the final output of noise suppression of the signal $s''(n)$.

3.12 Application Programming Interface

The following are the steps to utilize the noise suppression library:

(a) *Memory Allocation*: Allocate the memory for the input and output buffer placed in X memory and aligned to the 4- byte boundary.
- The size of the array should vary according to the user-defined noise suppression size of the frame (NS_FRAME).

(b) Every audio channel requires its own state holder in order to be processed therefore there is a need to allocate the memory for the noise suppression algorithm in the form of integer array in X memory, which is aligned at an address boundary of 4 bytes.
- Size of an array is specified by NS_XSTATE_MEM_SIZE_INT.

(c) *Allocate the scratch memory*: X and Y for noise suppression algorithm, which can be shared by multiple audio channels. The Y scratch memory is aligned at an address boundary of 512 bytes in the form of an integer array in Y memory, whereas X scratch memory is aligned at an address boundary of 4 bytes in the form of an integer array in X memory.

(d) *Initializing*: Use NS_init to initialize the noise suppression algorithm for an individual audio channel.

(e) *Suppressing noise*: Use NS_apply to remove the noise from the audio frame. Even if the requirement is not to suppress the noise in an audio frame the function is being called with the enable parameter set to false, that is NS_FALSE so that the noise suppression algorithm continues to adapt to the noise in the audio frames and therefore the audio is not affected.

3.13 Resource Requirements

To support dsPIC33, the following resources, including a program memory, data memory, dynamic usage, and computational speed, are required (see Tables 3.1–3.4) [13].

The total program memory is 174,590 bytes and out of which 8058 have been used by a noise suppression algorithm that is 4.63 percentage and system have total data memory of 16,384 bytes out of which 1284 have been used by noise suppression algorithm that is 7.83%. This implies that the proposed noise suppression algorithm uses <20% of the available system resources. Therefore the VAD, along with noise suppression, can be integrated with other low-resource, real-time applications, which can utilize the subband energies provided by the noise suppression algorithm.

3.14 Microchip PIC Programmer

Microchip's MPLAB IDE includes user-friendly features for the Microchip PIC programmers. Researchers and application developers can easily develop code for the dsPIC33. Various code libraries are available for many digital filter and voice codec applications.

TABLE 3.1 Program Memory

Type	Size (bytes)	Sections
Code	6675	.libns
Tables	1383	.const
Total	8058	–

TABLE 3.2 Data Memory

Functions	Size (bytes)	Alignment	Sections
nsStateMemX	768	4	X data memory
scratchmemX	100	4	X data memory
scratchmemY	256	512	Y data memory
signalIn	160	4	X data memory
Total	1284	–	–

TABLE 3.3 Dynamic Usage

Sections	Size (bytes)
Heap	0
Stack	<300

TABLE 3.4 Computational Speed

Functions	MIPS	Call Frequency
NS_init()	<0.5	Once
NS_apply()	3.6	10 ms

The following main components of the dsPIC board are shown in Fig. 3.4.

(a) *USB*: It provides the power to the system and a bidirectional communication between PC and the hardware circuit.

(b) *Linear voltage regulator (MCP1727)*: It's a 3.3 V linear regulator. It regulates the unregulated voltage of 5 V of PC to 3.3 V and supplies the power to the circuit.

(c) *Status LED for debug*: It indicates that a successful communication has been established between the hardware circuit and MPLAB IDE software.

(d) *Status LED for system power*: It indicated that the hardware is being powered by the USB.

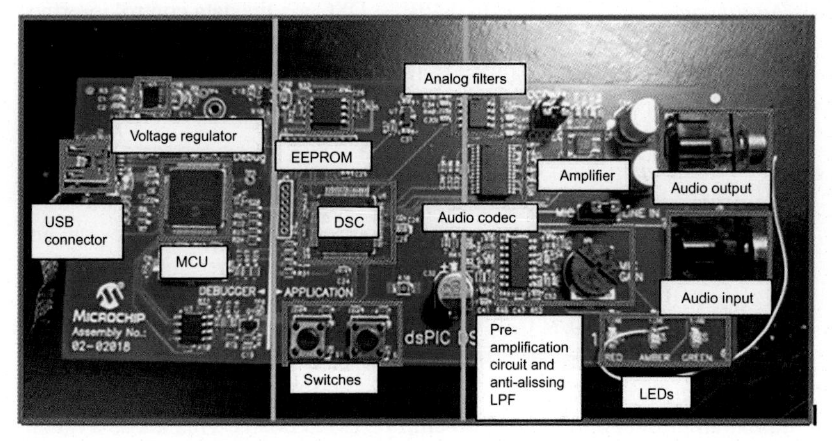

FIG. 3.4 Components of the dsPIC board.

(e) *PIC18F67J50 MCU*: This microcontroller controls the programming and debugging operations of the target device that is the dsPIC33FJ256GP506 DSC.

(f) *PNP transistor switch*: Through high-side switching, it provides the target power to the dsPIC33FJ256GP506 DSC, which acts as an ancillary circuitry, and this is controlled by the MCU through its debugging and programming operations.

(g) *Serial EEPROM (25LC010A)*: It acts as nonvolatile parameter storage for the microcontroller.

3.15 Audio Components

(a) *Flash memory*: This is a serial flash memory chip, which is provided by the regulated power supply by the linear voltage regulator. It provides the correct amount of current for the flash programming operation.

(b) *Digital signal control*: The DSC dsPIC33F256GP506 features 16kB RAM as well as 256kB of program flash, which helps in computation and as a processing resource for the development of an application. The algorithm/application can use on-chip FRC or externally provided 12MHz clock source.

(c) *PWM low pass filter (LPF)*: Pulse width modulation LPF demodulates the PWM signal, which is the output of the output compare module on the dsPIC33 DSC. The filter is of the fourth order, which uses two operational amplifiers on the quad op-amp IC (MCP6022). The jumpers provide us with the selection of the output. It determines whether the input is coming from the audio codec or the PWM LPFs to provide the input to the headphone amplifiers.

(d) *Audio codec*: It interfaces with dsPIC33 DSC via the data control interface module (DCI) and control interface bus (I2C). The output of the microphone amplifier is AC coupled to the codec. Wolfson WM8510 codec is being used which uses 12 MHz clock signal provided by the microcontroller.

(e) *Headphone amplifier*: For the study, a National Semiconductor stereo amplifier LM481170-Mw, along with the digital volume control was used. Logic levels are applied through the input/output ports of the device, and UP/DOWN pins of LM481170-Mw to the clock. The gain increases or decreases by 3 dB whenever the CLK line goes high depending on the logic level of the UP/DOWN line. There are 16 discrete gain settings, therefore, the gain can be adjusted over a range of 12 to −33 dB.

(f) *Line/microphone preamplifier*: This preamplifier uses one of the operational amplifiers on the quad op-amp IC (MCP6024). This act as a noninverting AC amplifier and the output is biased at 1.65 V. The potentiometer controls the gain of the amplifier. The gain is calculated using

$$Gain = \frac{1 + (R_{56} + R_{50})}{R_{44}} \tag{3.12}$$

(g) *Antialiasing LPF*: This LPF uses three operational amplifiers on the quad op-amp IC. The antialiasing LPF of the sixth order is being used by the output of the line/microphone preamplifier to filter the signal. It is a Sallen key structure and provides a cut off frequency of 3300 Hz.

3.16 VAD and Background Noise Reduction Techniques

The function of noise suppression is primarily based on the frequency domain algorithm. On each 10 ms frame of data, a FFT is applied to analyze the frequency components of a signal. Afterward, the algorithm of the VAD method helps us to determine whether the current frame is a noise or an original speech. Thereafter, the noise suppression algorithm maintains the database of the noise and updates itself whenever a noisy frame is detected by the VAD algorithm. Depending on the amount of noise in a particular frequency band, the input signal is then scaled in that frequency band, thereby causing the significant reduction of noise in the resultant signal. Thus this algorithm does not need an external reference for noise as it automatically adapts to the noise level and changes in nature. VAD helps in reducing the cochannel interference, power consumption in portable equipment, increasing number of radio channels, increasing the system capacity, enhancing speech coding quality, and avoiding unnecessary transmission of nonspeech signals.

Background noise results in deteriorating human-machine interaction in communication systems. These noises can be crowd noise, machinery, aircraft, background speech, etc. Even in a low SNR environment, the VAD algorithm should be fast enough to differentiate between speech and noise. Generally, the stability of human speech is around 20–30ms and VAD uses the processing frame-wise. The aim is to perform real-time background noise reduction in embedded systems and to implement it in low-power DSP.

3.17 Results and Discussion

Output SNR values in dB, when input speech SNR is closest to zero (Fig. 3.5). The selected experimental values of VAD hangover, noise factor, and noise suppression level are 6, 2, and 15, respectively. The output SNR value for babble noise without VAD and a noise suppression algorithm is 0.23dB. The value increased when VAD and noise suppression was implemented. The measured value was 6.44dB, which is a significant improvement in the speech signal. Experimental results were also improved when car noise, street noise, and airport noise were used as background noise. The value of SNR without VAD and a noise suppression algorithm for car noise, street noise and airport noise were 0.99, 0.18, and 0.22dB, respectively. The VAD and noise suppression algorithm were 5.54, 19.51, and 4.67dB, respectively.

Waveforms for the speech when SNR of the input signal with various noise when input speech SNR is 0dB are shown in the Fig. 3.6. The figure shows waveforms of the speech improvement by suppression of street noise that is achieved (a) without the VAD and noise suppression algorithm (noisy input signal), (b) with VAD and noise suppression algorithm.

If we increase the noise suppression value it will suppress the background noise, but along with it, the audio signal is, to some extent, attenuated. If a large value of hangover is applied, the algorithm will take much more to time to adapt itself to changes in the level of noise. On the other hand, if hangover of fewer frames is applied, the algorithm will adapt noise level changes faster, but it also

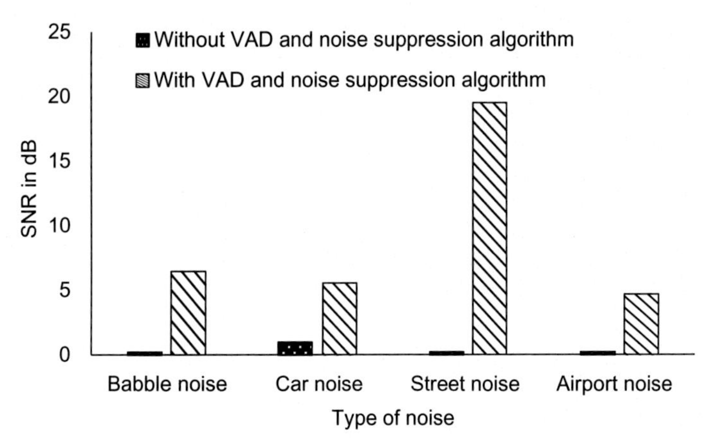

FIG. 3.5 Output SNR values in dB, when input speech SNR is closest to 0dB.

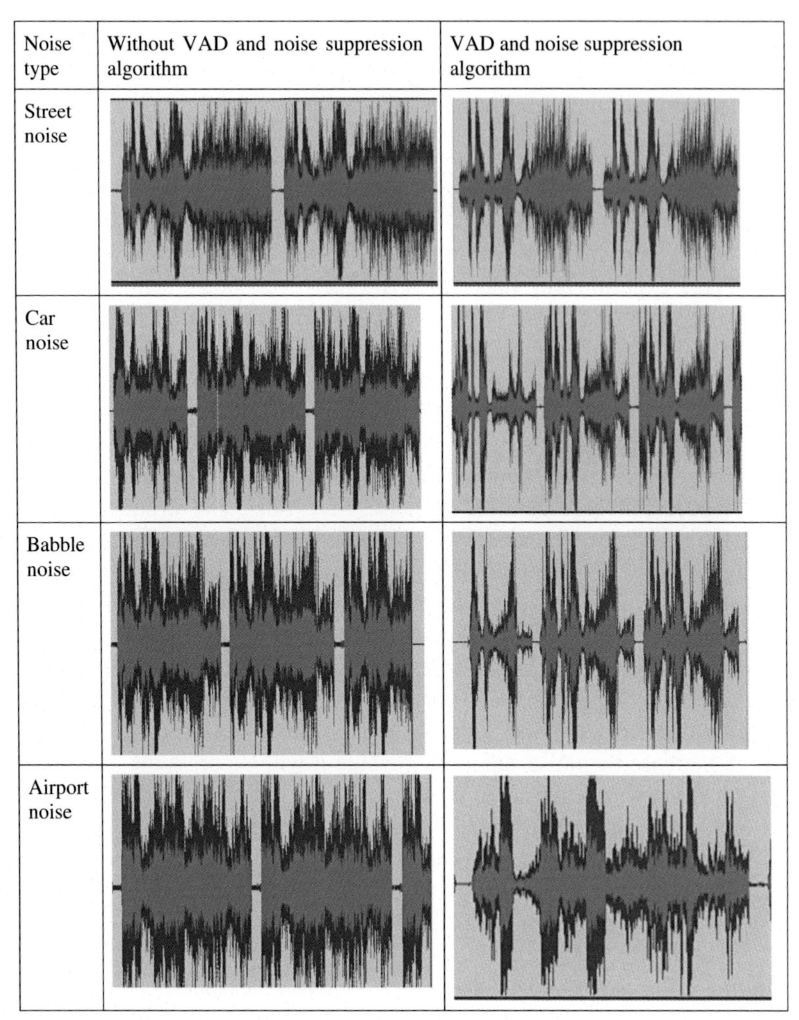

FIG. 3.6 Waveform for various noises. *Note*: All waveforms shown above have *X*-axis as amplitude and *Y*-axis as frames of the speech.

considers some of the speech signal frames as noisy frames. Signals that have noise levels greater than audio signals should have a larger noise factor, which is used by VAD to determine speech from noise. But, by doing so, it will flag more speech signals as noisy frames.

3.18 Conclusion and Discussion

For the algorithm development, MPLAB IDE software program use computer and developing applications for DSCs and Microchip microcontrollers. The implemented mechanism receives a noise-corrupted speech signal, suppresses the noise in the sampled signal, and plays the noise-suppressed signal. Military communication includes all aspects of communications of secured information and lot of advancement in the electronic age and the algorithm is suitable for real-time implementation in radio speech communication for military applications. Experimental results were taken when babble noise, car noise, street noise, and airport noise were used as background noise. The value of SNR values with various noise was significantly increased when VAD and noise suppression algorithm was used.

References

[1] M. Price, J. Glass, A.P. Chandrakasan, A low-power speech recognizer and voice activity detector using deep neural networks, IEEE J. Solid State Circuits 53 (2018) 66–75.

[2] N. Xu, C. Wang, J. Bao, Voice activity detection using entropy-based method, in: Proceedings of 9th International Conference on Signal Processing and Communication Systems, IEEE, Cairns, Australia, 2015. https://doi.org/10.1109/ICSPCS.2015.7391751.

[3] N. Dey, A.S. Ashour, Direction of Arrival Estimation and Localization of Multi-Speech Sources, Springer, 2018. https://doi.org/10.1007/978-3-319-73059-2.

[4] N. Dey, A.S. Ashour, W.S. Mohamed, N.G. Nguyen, Acoustic wave technology, in: Acoustic Sensors for Biomedical Applications, SpringerBriefs in Speech Technology (Studies in Speech Signal Processing, Natural Language Understanding, and Machine Learning), Springer, Cham, 2019, pp. 21–31.

[5] N. Dey, A.S. Ashour, W.S. Mohamed, N.G. Nguyen, Acoustic Sensors in Biomedical Applications, in: Acoustic Sensors for Biomedical Applications, SpringerBriefs in Speech Technology (Studies in Speech Signal Processing, Natural Language Understanding, and Machine Learning), Springer, Cham, 2019, pp. 43–47.

[6] A.M. Johansson, E.A. Lehmann, S. Nordholm, Real-time implementation of a particle filter with integrated voice activity detector for acoustic speaker tracking, in: IEEE Asia-Pacific Conference on Circuits and Systems, Proceedings, APCCAS, 2006, pp. 1004–1007, https://doi.org/10.1109/APCCAS.2006.342257.

[7] R.A. Lara-Cueva, A.S. Moreno, J.C. Larco, D.S. Benitez, Real-Time Seismic Event Detection Using Voice Activity Detection Techniques, IEEE J. Sel. Top. Appl. Earth Obs. Remote Sens. 9 (12) (2016) 5533–5542, https://doi.org/10.1109/JSTARS.2016.2605061.

[8] N. Lezzoum, G. Gagnon, J. Voix, Voice Activity Detection System for Smart Earphones, IEEE Trans. Consum. Electron. 60 (4) (2014) 737–744, https://doi.org/10.1109/TCE.2014.7027350.

[9] A. Sehgal, N. Kehtarnavaz, A convolutional neural network smartphone app for real-time voice activity detection, IEEE Access 6 (2018) 9017–9026, https://doi.org/10.1109/ACCESS.2018.2800728.

[10] P.Y. Shih, P.C. Lin, J.F. Wang, Y.Z. Chen, Improving real-time voice activity detection for perceptual robotic control in noisy environment, in: IEEE Region 10 Annual International Conference, Proceedings/TENCON, 2011, pp. 1040–1044, https://doi.org/10.1109/TENCON.2011.6129269.

[11] dsPIC33F Family Data Sheet, High-Performance, 16-bit Digital Signal Controllers, Microchip Technical Literature, Microchip Technology Inc., 2006. Retrieved from, http://ww1.microchip.com/downloads/en/DeviceDoc/70165d.pdf. Accessed 15 February 2018.

[12] New Microchip dsPIC33 Digital Signal Controller Family, http://www.microcontroller.com/news/microchip_dsPIC33.asp, 2005.

[13] dsPIC® DSC Noise Suppression Library User's Guide, 2004–2011 Microchip Technology Inc. http://ww1.microchip.com/downloads/en/DeviceDoc/DS-70133E.pdf

Further Reading

M. Asadullah, A. Raza, An overview of home automation systems, in: Proceedings of 2nd International Conference on Robotics and Artificial Intelligence, IEEE Press, Rawalpindi, Pakistan, 2016.

S. Soumya, M. Chavali, S. Gupta, Internet of things based home automation system, in: Proceedings of IEEE International Conference on Recent Trends in Electronics, Information & Communication Technology, IEEE Press, Bangalore, India, 2017.

Chapter 4

Disambiguating Conflicting Classification Results in AVSR

Gonzalo D. Sad, Lucas D. Terissi and Juan C. Gómez
Laboratory for System Dynamics and Signal Processing, Universidad Nacional de Rosario, CIFASIS-CONICET, Rosario, Argentina

4.1 Introduction

Besides the acoustic signal, it is understood that the visual information during speech such as facial expressions, hand gestures, and body posture contributes significantly to the intelligibility of the message being transmitted, and to the perception of the actual meaning of the message [1]. In addition, as pointed out in a recent survey about the interaction between gesture and speech [2], the parallel use of these modalities gives the listener access to complementary information not present in the acoustic signal by itself. In recent years, the study of human communication has benefited from the increasing number of multimodal corpora available to researchers in this field. Significant research effort has been devoted to the development of audio-visual speech recognition (AVSR) systems where the acoustic and visual information (mouth movements, facial gestures, etc.) during speech are taken into account [3–5].

In recent years, several methods, algorithms and models, have been developed to perform speech recognition by fusioning audio and visual information. One of the first methods presented in the literature (and one still widely used) is hidden Markov models (HMMs), which provide results that are particularly useful because it can handle time series (like speech signals) in a very efficient way [6–14]. Other classical methods from the machine learning field have also been implemented, like artificial neural networks (ANN) [15], *k*-nearest neighbors (*k*-NN) algorithms [16], matching methods utilizing dynamic programming, adaptive boosting classifiers (AdaBoost) [17], support vector machine (SVM) [18], linear discriminant analysis [19], random forests (RFs) [20], etc. Most of these methods cannot handle time series, so a preprocessing step is required to perform a normalization procedure. In recent years, newer and more sophisticated methods, like restricted Boltzmann machines (RBM) [21,22], deep learning (DL) [4,23–27] and sparse coding [28–31], have proven to be

Intelligent Speech Signal Processing. https://doi.org/10.1016/B978-0-12-818130-0.00004-0

suitable for speech recognition tasks by improving the recognition rates obtained with traditional methods like HMMs. This improvement is at the expense of increasing the complexity of the system and the requirement of much more data in the training steps.

All the above mentioned methods, can be classified either as generative models or discriminative models. Given an observable variable X and a target variable Y, a generative model is a statistical model of the joint probability distribution $P(X,Y)$, while a discriminative model is a model of the conditional probability of the target Y, given an observation x, symbolically, $P(Y|X=x)$. In speech recognition tasks, the generative models (HMMs, RBM, etc.) are formed using one model for each class, that is, one model for each phoneme or word that compose the dictionary of the problem being analyzed. In the training step, each model is trained using examples of the class to be represented. On the other hand, when a discriminative model is used (RF, SVM, *k*-NN, ANN, etc.) for speech recognition tasks, only one classifier is formed, which internally defines all the classes forming the dictionary.

It has been found that improved results can be obtained in recognition tasks by using a combination of models [32–34]. These improvements can be obtained if the combined models are complementary, in the sense that they deliver different classification results. There are several reasons why a combination of systems can outperform a single system, and it is necessary to consider the limitations of using a single system. There are several ways to improve the performance of a single system. First, the underlying model parameters can be improved, perhaps by ignoring simple modeling assumptions, to better model the data. Second, the models can be made more complex by increasing the number of parameters in the system so they have more power to model the data. However, it is well known that increasing the number of parameters can lead to overtraining and a lack of generalization. To address this problem, more training data is typically added. Unfortunately, increasing the amount of training data is not always the most efficient way to improve results since it is expensive to accurately transcribe a large database, and the gains achieved from incorporating more training data become increasingly smaller. Finally, improved training algorithms, such as discriminative training, can be used to improve the underlying models. Again, sophisticated algorithms can lead to issues with overtraining, and there is a limit on how much improvement they can achieve in practice. Dietterich [35] suggests three theoretical reasons why an ensemble or combination of classifiers may perform better than just one classifier alone:

- *Lack of training data:* With limited training data, the estimated models will only be an approximation to the true underlying data distribution. The average of a number of different estimates may be closer to the true model than any of the individual estimates.
- *Limitations of the optimization algorithm:* In practice, the optimization algorithm is normally only able to find a local optimum for parameter

estimates, except in simple situations. Again, an average of multiple estimates may be closer to the global optimum than a single estimate.

- *Representational issues:* It may be the case that the model has limited capability and is unable to represent the true data distribution. A combination of multiple models may be able to represent distributions, which cannot be modeled by just one model.

Ensembles of models for machine learning tasks have been a focus of research for several years [35], and approaches to building complementary systems fall broadly into two categories. First, diverse systems are somehow generated, and it is hoped they will have different errors [36–39]. Examples of this approach include altering the training algorithm, frontend, and covariance modeling. Second, complementary systems can be explicitly trained [40–44]. This can be done either by iteratively building multiple complementary systems, or by training the complementary systems in parallel. This approach usually yields an overfitted model, due to the excessive complexity injected in the training stage. Previous work in AVSR has mainly relied on an ad hoc approach to finding complementary systems. Multiple systems are built, and those that perform well in combination are selected. Generating complementary systems is a current research topic in AVSR, and for machine learning in general.

In this chapter, a novel scheme for speech classification tasks based on the combination of traditional and complementary models in a cascade structure, is proposed to improve recognition rates. The proposed scheme can be implemented with different models, viz., generative models like HMM or RBM and discriminative models like RF, SVM, AdaBoost, ANN, k-NN, etc. There are no restrictions about the kind of speech features employed as inputs of the proposed method. For instance, it can be used for lip-reading tasks by considering visual features as inputs, or for audio speech recognition using acoustic features as inputs, and also for audio-visual speech recognition where the inputs are acoustic and visual features previously fusionated (early integration). Given a model or classifier (HMM, RF, SVM, etc.), the corresponding complementary model proposed in this chapter is generated using the same training procedure as in the original model or classifier, but defining a new set of classes for the training examples, aiming to detect absence of a class. In the complementary models, the ith class is formed using all the instances in the vocabulary except those corresponding to class i. For instance, let us consider a vocabulary composed by three classes, C_1, C_2, and C_3, the new classes are defined as AC_1, which contains all the examples of the classes C_2 and C_3, AC_2, which contains all the examples of the classes C_1 and C_3, and AC_3, which contains all the examples of the classes C_1 and C_2.

Given an example to be classified, the first step in the proposed cascade of classifiers is carried out by making an initial classification by evaluating the traditional model. From the output probabilities of this traditional classifier, only the M most likely classes are preselected, and the rest are discarded as possible solutions. If any pair of conflicting classes in this group of M classes doesn't exist, the recognized class is defined by the highest-ranked class from this preselected group of classes. Otherwise, if there exists any conflicting classes, a second step is carried out where the complementary models are evaluated and their outputs probabilities are used to determine the recognized class.

The objective of this work is to disambiguate conflicting classes in order to improve the resulting recognition rates. For this purpose, a cascade scheme is proposed considering the novel concept of complementary models.

The performance of the proposed system is evaluated with four different models, viz., one generative model (HMM) and three discriminative models (RF, SVM, and AdaBoost). In order to evaluate the performance of the proposed system, different experiments are carried out over two public databases, viz., AVLetters database [45], Carnegie Mellon University (CMU) database [46], and one database compiled by the authors, hereafter referred to as AV-UNR database. As mentioned before, the proposed method has no restrictions about the kind of input features, so experiments with different input features are carried out, viz., audio features, visual features (lip-reading) and fused audio-visual features (early integration). Since in most of real applications the acoustic channel is affected by noise, the performance of the proposed cascade of classifiers is evaluated injecting different kinds of noise in the acoustic information, viz., Gaussian noise and additive babble noise [47]. Promising results are achieved with the proposed method, obtaining significant improvements in all the considered scenarios. Also, better results than other methods reported in the literature are achieved in most of the cases over the two public databases. Preliminary results of this work have been presented in the conference papers [48,49], where the system was evaluated for word classification considering only HMMs models with audio and audio-visual input features and an RF classifier with audio, visual, and audio-visual input features, respectively. In this chapter, a complete analysis of the proposed classification scheme is described, considering audio, visual, and fused audio-visual information, and implemented with four different models and classifiers (HMM, SVM, RF, and AdaBoost). Additionally, the proposed scheme is extended and evaluated over three different databases.

The remainder of this chapter is organized as follows. Section 4.2 presents a preliminary analysis of the class confusability problem and Section 4.3 introduces and explains the classification based on absent classes. In Section 4.4, a description of the proposed system is given. The databases used for the experiments are described in Section 4.5, and the experimental results and the performance of the proposed strategy is analyzed in Section 4.6. Finally, some concluding remarks are given in Section 4.7.

4.2 Detection of Conflicting Classes

In speech recognition tasks, usually the units to classify or recognize are phonemes or words, that is, each of these units are represented by a class within the classifier. Generally, the errors made by the classifier are not completely random, that is, it can be observed that usual errors correspond to certain classes or group of classes. This source of error is known as class confusability or intra-class confusion. For example, in visual speech recognition tasks, the majority of errors or confusion is between sequences consisting of the same visemes (a viseme is defined as the smallest visibly distinguishable unit of speech). If we would like to recognize the alphabetical letters [A-Z], we would see that the most frequent errors would be between the utterances corresponding to the letters B and P, which are composed of phonemes [B+IY] and [P+IY], respectively. If we map the phonemes to their corresponding visemes, we can see that these words are visually the same and composed of the visemes /p+iy/. Similarly, the words C, D and T, which are composed of phonemes [S+IY], [D+IY] and [T+IY], respectively, are composed of the same sequence of visemes, /t+iy/, and are also a major source of errors.

This kind of behavior is easily observable in the confusion matrix obtained from the classification results. If the problem at hand has p classes, the resulting confusion matrix is a matrix with p rows and p columns, with the rows representing the classifier decision and the columns representing the true classes. If the classifier makes no errors, the values of off-diagonal entries will be zeros, but if the problem at hand suffers from class confusability or intra-class errors, there will be a few off-diagonal elements with high values while the rest will have very low values. So, looking at the position (row and column) of these high value off-diagonal elements, we can find the most instances of class confusability. In this chapter, the P most conflicting classes will be determined by taking the position (row and column) of the P highest values of the off-diagonal elements of the confusion matrix.

4.3 Complementary Models for Classification

The different models and classifiers proposed for speech recognition tasks can be classified into two main groups, namely, generative models (for instance HMM, RBM, etc.) and discriminative models (RF, SVM, boosting methods, ANN, etc.). A single model is usually used to represent each word or phoneme in a given class when generative models are employed. Each of these models are trained with examples of each particular class. Given the observation sequence to be recognized (O), the models are evaluated and the one given the highest probability of observation is the one that determines the recognized class. When discriminative models are used, a single classifier representing all the classes in the dictionary is trained. Given an observation sequence to be recognized (O), the model is evaluated and the probabilities for each class are obtained. The recognized class corresponds to the one that obtains the highest probability.

An alternative to using the data in the training step to produce a new model is presented in this paper. In contrast to traditional classifiers, the aim of these models is to detect the absence of a class. For this reason, the classification technique will be referred to as the complementary model approach. The main idea is to detect the absence of a class by redefining classes, either internally within the model in the case of being discriminative, or individually in the case of being generative. Given a vocabulary composed by N classes, the complementary model is generated by redefining the original N classes. For each class i, a new class is defined, denoted as AC_i, using all the instances in the vocabulary except the corresponding to class i. That is,

$$\begin{cases} AC_i = I_N - \{C_i\} \quad & i = 1,2,\ldots,N \\ I_N = \{C_1, C_2, \ldots, C_N\} \end{cases} \tag{4.1}$$

where C_i is the original classes, I_N is the set representing the N original classes, $I_N - \{C_i\}$ is the set containing all the N original classes except the C_i class, and AC_i is the new classes. Fig. 4.1 schematically depicts how the new classes are formed for a vocabulary of $N=3$ classes.

Since the models are trained with the complementary classes AC_i, it would be reasonable to expect that the minimum value of the computed probabilities of an input observation sequence belonging to the ith class, will correspond to the ith class. In this way, the recognized word will correspond to the class given the minimum probability, since the model is detecting the absence of the ith class. The decision rule for the case of discriminative model λ can be expressed as

$$\begin{gathered} i = \underset{j}{\operatorname{argmin}}\ P\left(O|\lambda_{AC}, AC_j\right) \\ AC_j = I_N - \{C_j\} \\ I_N = \{C_1, C_2, \ldots, C_N\}, \end{gathered} \tag{4.2}$$

where i is the recognized class, λ_{AC} is the classifier that has been trained with the new set of classes (AC_j), which will hereafter be referred to as the *complementary discriminative model*, and $P(O \mid \lambda_{AC}, AC_j)$ indicates the probability of the new class AC_j, given the model λ_{AC} and the observation sequence O. For the case of generative model Ω, this decision rule is expressed by the following equation

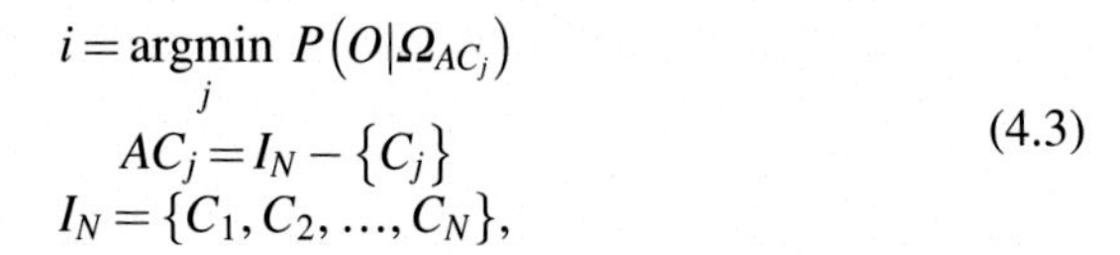

$$\begin{gathered} i = \underset{j}{\operatorname{argmin}}\ P\left(O|\Omega_{AC_j}\right) \\ AC_j = I_N - \{C_j\} \\ I_N = \{C_1, C_2, \ldots, C_N\}, \end{gathered} \tag{4.3}$$

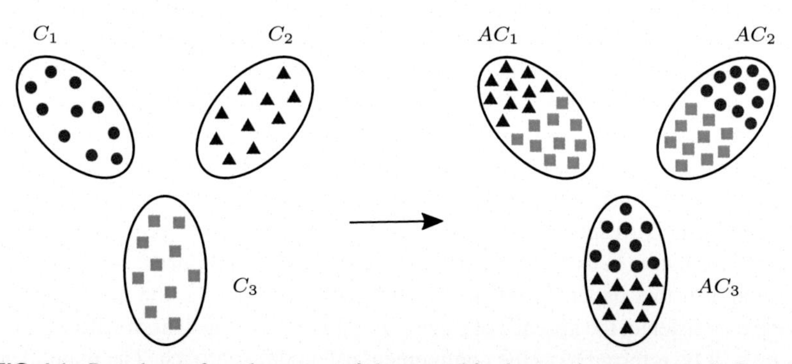

FIG. 4.1 Procedure to form the new set of classes used for the complementary models. Example for a vocabulary of size $N=3$, where $\{C_1, C_2, C_3\}$ are the original classes, and $\{AC_1, AC_2, AC_3\}$ are the new ones.

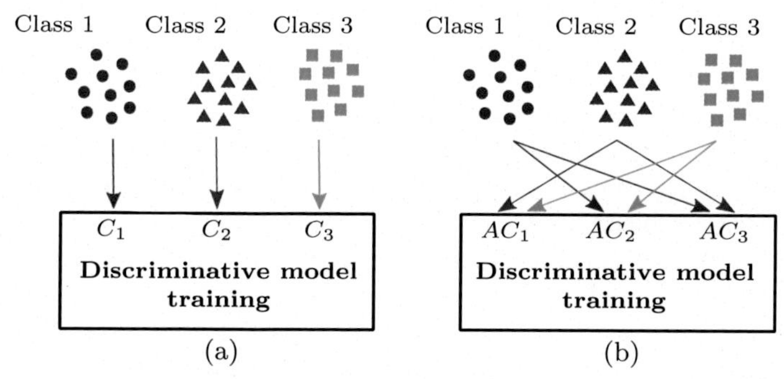

FIG. 4.2 Training procedure of the proposed complementary models for the discriminative models and a vocabulary of size $N=3$. (a) Traditional model with classes $\{C_1, C_2, C_3\}$. (b) Complementary model with the new set of classes $\{AC_1, AC_2, AC_3\}$.

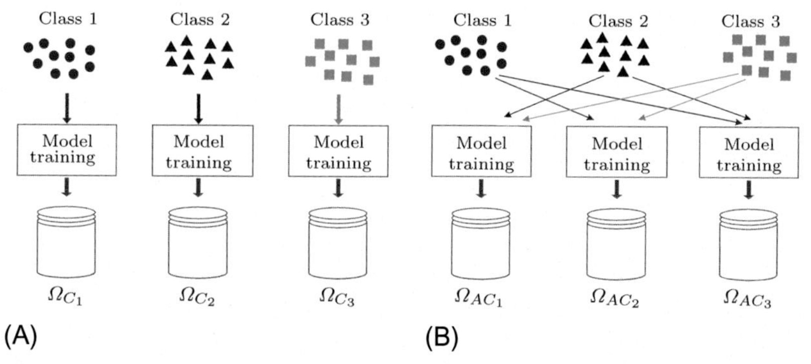

FIG. 4.3 Training procedure of the proposed complementary models for the case of generative models and a vocabulary of size $N=3$. (a) Traditional models $\{\Omega_{C_1}, \Omega_{C_2}, \Omega_{C_3}\}$, one for each class. (b) Complementary models $\{\Omega_{AC_1}, \Omega_{AC_2}, \Omega_{AC_3}\}$, one for each new class $\{AC_1, AC_2, AC_3\}$.

where i is the recognized class, Ω_{AC_j} is the model that has been trained with the new class AC_j, which will hereafter be referred to as the *complementary generative model*, and $P(O\,|\,\Omega_{AC_j})$ indicates the output probability of model Ω_{AC_j}, given the observation sequence O.

Fig. 4.2 schematically depicts the training procedure for a discriminative model λ and its corresponding complementary model λ_{AC}. In Fig. 4.3, the training procedure for a generative model Ω and its corresponding complementary model Ω_{AC}, is also shown. In both cases, the vocabulary has three different classes ($N=3$).

4.4 Proposed Cascade of Classifiers

It has been found that using a combination of systems for AVSR can outperform the use of a single system. For the combination to lead to improvements, the individual models must be complementary, that is, they must make different errors. The multiple diverse systems can be built in many ways, including the use of different frontends, injecting randomness, altering the model topology, using different training algorithms, etc. The complementary systems may correct errors made by previous systems, but can introduce new errors. However, if the systems make complementary errors, they should lead to improvements when combined.

In this section, a cascade classification structure is proposed to improve recognition rates. This scheme for speech classification is based on the combination of traditional and complementary models, and it could be implemented with both generative and discriminative models. The proposed combination scheme uses the complementary models to rescore only when it is believed that the traditional system is incorrect. Also, it can be employed with different kinds of input information, viz., audio, visual, or audio-visual information, indistinctly. Given a model (HMM, RF, SVM, ANN, etc.), the proposed cascade of classifiers is formed with the traditional model and the complementary version of the traditional model, based on the method proposed in Section 4.3. A schematic representation of the proposed speech classification scheme is depicted in Fig. 4.4.

In the training stage, the features obtained from the input audio-visual speech data are used to train the traditional model (λ or Ω). This model is used to detect the conflicting classes, based on the analysis of the confusion matrix described in Section 4.2. Based on this analysis, the complementary models are generated and trained. The test stage is composed by two steps. First, a classification based on the traditional model is performed, and the M most likely classes are preselected using the output probabilities. At this point, the possible solutions are reduced to these M classes. If none of these preselected classes

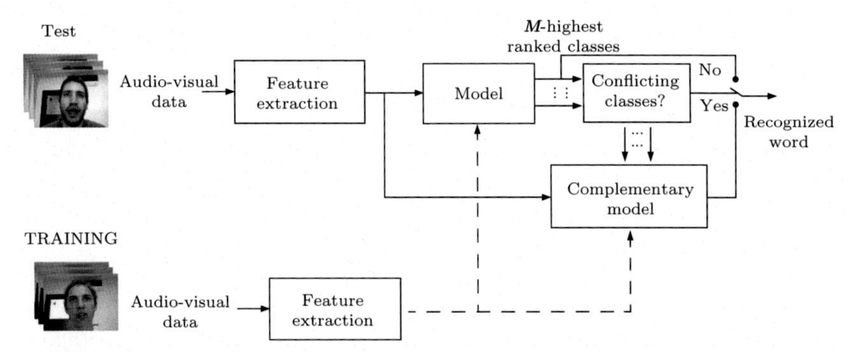

FIG. 4.4 Schematic representation of the proposed cascade of classifiers based on complementary models for audio-visual speech classification. The M-highest ranked classes are obtained by sorting the model's output probabilities.

are considered as conflicted ones (based on the analysis carried out in the training stage), the recognized word corresponds to the one with the highest probability. Otherwise, if any pair of these M classes is considered as conflicted, a second step is performed. In this step, the complementary model (λ_{AC} or Ω_{AC}) associated with these M classes is selected and used to determine the recognized class.

Fig. 4.5 schematically depicts the cascade of classifiers scheme proposed in this chapter, for the case of $M=3$ and implemented with a discriminative model λ. Given an observation sequence O, associated with the word to be recognized, the λ original model is evaluated, and its output probabilities ($P(O\,|\,\lambda, C_i)$, $i=1, 2, \ldots, N$) are ranked. The $M=3$ highest ranked values define the classes ($I_3=\{C_6, C_3, C_9\}$) to be used for selecting the complementary model λ_{AC} (which has been previously trained in the training stage), because there are conflicting classes (C_3 and C_6). In this case, the new classes defined inside the complementary model are: $AC_3=I_3-\{C_3\}$, which is trained with the training data corresponding to classes C_6 and C_9; $AC_6=I_3-\{C_6\}$, which is trained with the training data corresponding to classes C_3 and C_9; and $AC_9=I_3-\{C_9\}$, which is trained with the training data corresponding to classes C_3 and C_6. Finally, the complementary model defined with these new classes is evaluated and the recognized word corresponds to the new class AC_i that gives the minimum probability.

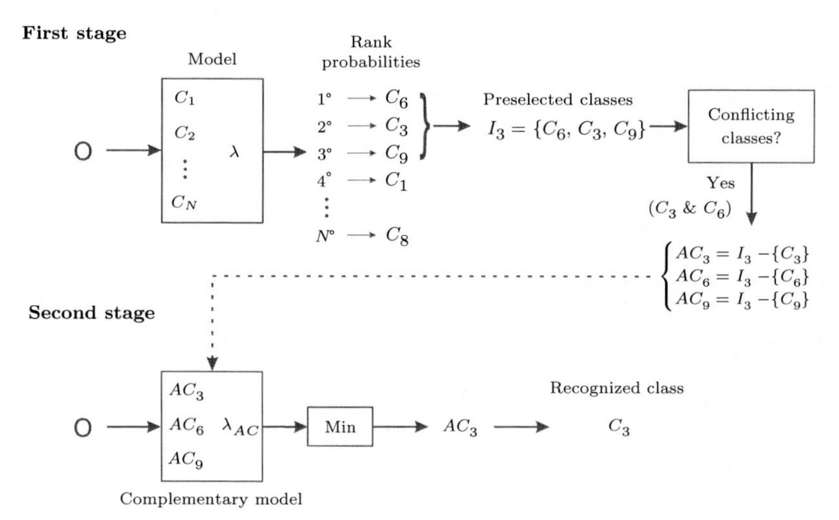

FIG. 4.5 Example of the proposed classifier combination strategy with $M=3$ and implemented with a discriminative model λ. In the second stage, the complementary model (λ_{AC}) uses the new set of classes: $\{AC_3, AC_6, AC_9\}$.

4.5 Audio-Visual Databases

The performance of the proposed classification scheme is evaluated over three different audio-visual databases, two well-known public ones, and another compiled by the authors of this chapter. Each database has its own visual speech representation. In particular, for two databases, visual features were computed using model-based methods, while image-based features were considered as visual information for the remaining one. Each database is described below.

4.5.1 AV-CMU Database

The AV-CMU database [46] consists in the recording of 10 speakers uttering a series of words. In this chapter, a subset of 10 words, corresponding to numbers from 1 to 10 is considered for the experiments. These numbers were pronounced 10 times by each speaker. Acoustic data was captured at 44.1 kHz, while visual data was captured at 30 frames per second. As depicted in Fig. 4.6A, the database provides the visual data through the position of the mouth and lips at each video frame. In this chapter, the parabolic lip contour model, proposed by Borgström and Alwan [10], is employed to represent the visual information during speech. This parabolic lip model, depicted in Fig. 4.6B, is fitted using the left (x_1, y_1) and right (x_2, y_2) corner positions of the mouth, and the heights of the openings of the upper (h_1) and lower (h_2) lips. Visual features are then represented by five parameters, namely, the focal parameters of the upper and lower parabolas, mouth's width and height, and the main angle of the bounding rectangle of the mouth.

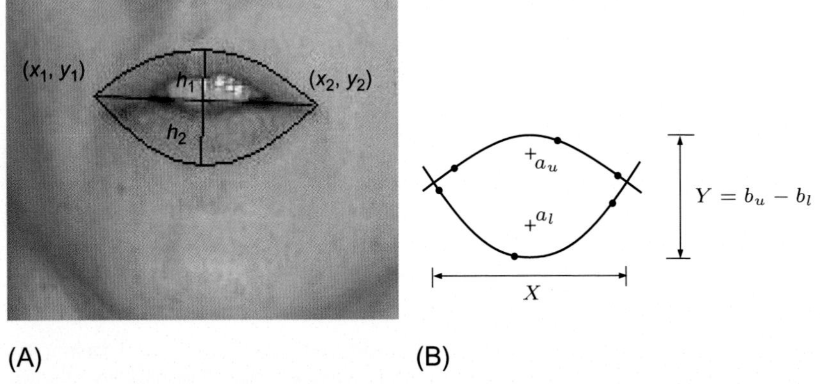

FIG. 4.6 AV-CMU database. (A) Visual data included in the database. (B) Parabolic lip contour model proposed by Borgström and Alwan [10].

4.5.2 AV-UNR Database

This database, compiled by the authors of this chapter, is composed by the utterances of a set of 10 words (corresponding to actions such as open, close, save, stop, etc.), performed in random order by 16 speakers. Each word was pronounced 20 times by each speaker, resulting in a total of 3200 utterances. Audio data was captured at 8 kHz, while visual data was captured at 60 frames per second with a resolution of 640×480 pixels. To represent the visual information during speech, the model-based method proposed in Terissi and Gómez [50] is employed. This method extracts mouth visual features using a simple 3D face model, namely Candide-3 [51]. As depicted in Fig. 4.7, at each video frame, visual information is represented by three parameters, viz., mouth height (v_H), mouth width (v_W), and area between lips (v_A).

4.5.3 AVLetters Database

The AVLetters database [45] is composed by the recording of the letters (A-Z) by 10 speakers (five males and five females). In these recordings, each speaker pronounced each letter three times, resulting in a total of 780 utterances. The original acoustic voice signals are not provided by this database. However, it includes the corresponding mel-frequency cepstral coefficients computed for each utterance. Visual data consists of mouth region images with a 80×60 pixels resolution, captured at 25 frames per second. Example images from the 10 speakers are included in Fig. 4.8. To represent the visual information during speech, the method based on local spatiotemporal descriptors proposed in [18] is employed in this chapter. This image-based method is applied directly over the image sequences of the AVLetters database. As a result of this method [18], each image of the database is represented by a feature vector of 1770 coefficients.

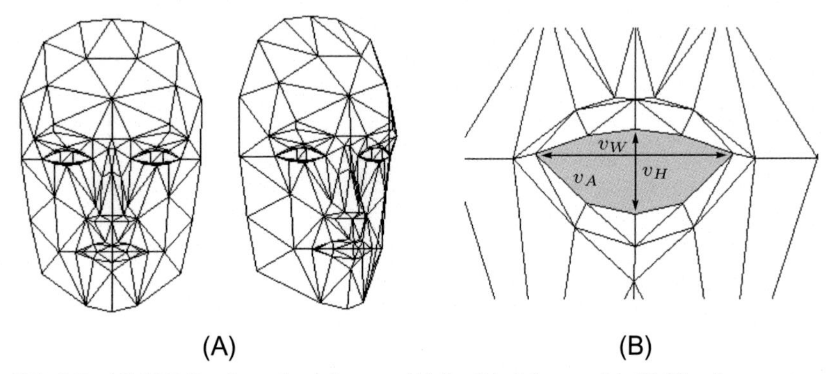

FIG. 4.7 AV-UNR Database visual features. (A) Candide-3 face model. (B) Visual parameters.

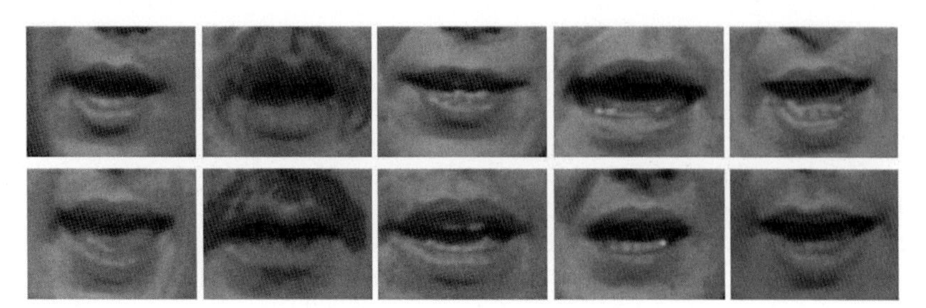

FIG. 4.8 Example images from 10 speakers of the AVLetters database.

4.6 Experimental Results

In order to evaluate the performance and robustness of the proposed method in this chapter, several experiments are carried out. As mentioned before, the cascade of classifiers proposed in Section 4.4 can handle different kinds of input features, so experiments with audio features (audio speech recognition), visual features (visual speech recognition or lip-reading), and fused audio-visual features (AVSR) are performed. Regarding the models and classifiers used to implement each stage of the system described in Section 4.4, four different cases are evaluated: HMM, RF, SVM and AdaBoost. That is, a total of three discriminative models and one generative model are analyzed. The three databases reported in Section 4.5 are employed to carry out all the experiments described above.

Different factors exist in real applications that usually are not taken into account in the training stage of speech recognition systems. In most cases, this is due to the complexity of the models or the lack of data representing all the possible scenarios in real situations. For example, visual occlusion in visual or AVSR, and acoustic noise affecting the audio information in audio or AVSR. Since the presence of acoustic noise is one of the most common sources of errors in speech recognition systems, experiments with noisy audio conditions are performed. Additive Gaussian and additive babble noise [47] is injected in the acoustic channel for all the experiments where acoustic information is employed to extract the input features to the proposed system, in order to evaluate the robustness of the proposed cascade of classifiers. The NOISEX-92 database [52] is employed to extract samples of babble noise, which is one of the most challenging noise conditions because it is generated by speech from other speakers interfering with the original speech to be recognized. This acoustic noise robustness analysis is carried out using clean audio information for the training stage and noise corrupted audio information (with signal-to-noise ratios (SNRs) ranging from −10 dB to 40 dB) for the testing stage of the proposed system.

In all the scenarios described above, a D-fold cross-validation procedure is carried out to obtain statistically significant recognition rates. Particularly, speaker independent evaluations are performed, using only one speaker in the testing stage and the remaining ones in the training stage, resulting in a 10-fold for AV-CMU database and 16-fold for AV-UNR database. In order to fairly compare the results obtained in this work over the AVLetters database with others methods reported in the literature, the evaluation protocol used by the authors in [18,21,22,45] is employed. The first two utterances of each speaker are used in the training stage, and the third utterance of each speaker is used in the testing stage. Unlike the case of AV-CMU and AV-UNR databases, for AVLetters database the evaluation is speaker dependent.

A frame wise analysis over the acoustic signal is performed and the first eleven Mel-Cepstral coefficients, discarding the zero-order coefficient which is related to the energy of the input signal, are extracted frame by frame. Also, their associated first and second time derivatives are calculated, which gives an audio feature vector composed by 33 parameters for each frame. In the AVSR experiments, the same frame-wise analysis is performed, partitioning the acoustic signal in frames. The frame rate is defined by the video channel frame rate. For each frame, the audio-visual feature vector is composed by the concatenation of the corresponding acoustic and visual features. This results in a fused audio-visual feature vector composed by 38 and 36 parameters for the AV-CMU and AV-UNR databases, respectively. Since the original recordings of the acoustic signals from AVLetters database are not available, experiments under noisy acoustic conditions for audio and AVSR scenarios could not be performed.

As described in Section 4.1, there are many models and classifiers proposed in the literature to perform classification and recognition tasks. The vast majority, especially when using discriminative models, cannot handle variable length input data, as is the case of speech recognition systems, due to its time varying nature. Each utterance, even of the same word, is highly unlikely to have the same temporal duration. This leads to the need of some kind of normalization process to obtain fixed-length input data representation. To this end, the proposed method in [20] based on a wavelet feature extraction technique, is employed in all the experiments with RF, SVM, and ADA classifiers. This method requires three parameters to be defined, viz., the mother wavelet, the resolution level for the approximation, and the normalized length of the resampled time functions. In this work, the wavelet resolution level was set to three, the normalized length of the resampled time functions was set to 256, and the Daubechies 4 wavelet (db4) was chosen as the mother wavelet.

To disambiguate conflicting classification results, the conflicted classes must be recognized as a first step. The method proposed in Section 4.2 is employed, evaluating the classifier (HMM, RF, SVM or ADA) and selecting the four most conflicting pair of classes ($P=4$) based on the analysis of the confusion matrix obtained. For each scenario, this evaluation is carried out randomly partitioning the whole data of each database in a training set (70%) and a test set (30%). These four pairs of conflicted classes are used in the second stage of the proposed system in Fig. 4.4 to determine if there are conflicting results for the observation being processed in the testing stage. The complementary models represented in Fig. 4.4 were implemented as 3-class complementary ($M=3$).

4.6.1 Hidden Markov Models

In this section, the method proposed in Section 4.4 for generative models is applied to traditional HMMs [53]. In the first stage, the classifiers are implemented with HMM while in the second stage the complementary models are implemented with Gaussian mixture models (GMMs), namely complementary Gaussian mixture models (CGMMs). The classical N-state, left-to-right structure with continuous symbol observation and a linear combination of M_h Gaussian distributions as representation for each state, was employed for the HMMs implementation. The GMMs used to implement the CGMMs were modeled by a linear combination of M_g Gaussian distributions with continuous observations. The expectation-maximization algorithm was employed for training both, the HMMs and the GMMs.

In this case, there are three tuning parameters of the proposed system, namely, the number of states (N) and Gaussian distributions (M_h) of the HMMs, and the number of Gaussian distributions (M_g) of the CGMMs, which are optimized via exhaustive search. Several experiments were carried out using values of N in the range of 1–20, M_h from 1 to 30, and M_g from 8 to 256, looking for the combination giving the best recognition results.

The recognition rates obtained at different SNRs over the AV-CMU and AV-UNR databases, using audio-only information and fused audio-visual information are depicted in Fig. 4.9, for HMM and the proposed cascade of classifiers (C-HMM). It is clear that, for both databases, both types of inputs, and both kinds of acoustic noise, the proposed scheme (C-HMM) performs better than the ones based on HMMs. This is more apparent at low- and middle-range SNRs. In some cases, there are very significant improvements, for example in Fig. 4.9F, for AV-CMU database using audio-only information and Gaussian noise, reaching almost 33% of improvement for SNR = 10 dB. The results for the lip-reading scenario for AV-UNR, AV-CMU and AVLetters databases, are shown in Table 4.1. It shows the use of the proposed cascade of classifiers scheme (C-HMM) improves the recognition rates for all databases.

4.6.2 Random Forest

In this section, the method proposed in Section 4.4 for discriminative models is applied to RF [54]. Both stages of the proposed cascade of classifiers are implemented with RF. The complementary model based on RF will be hereafter referred to as complementary random forest (CRF).

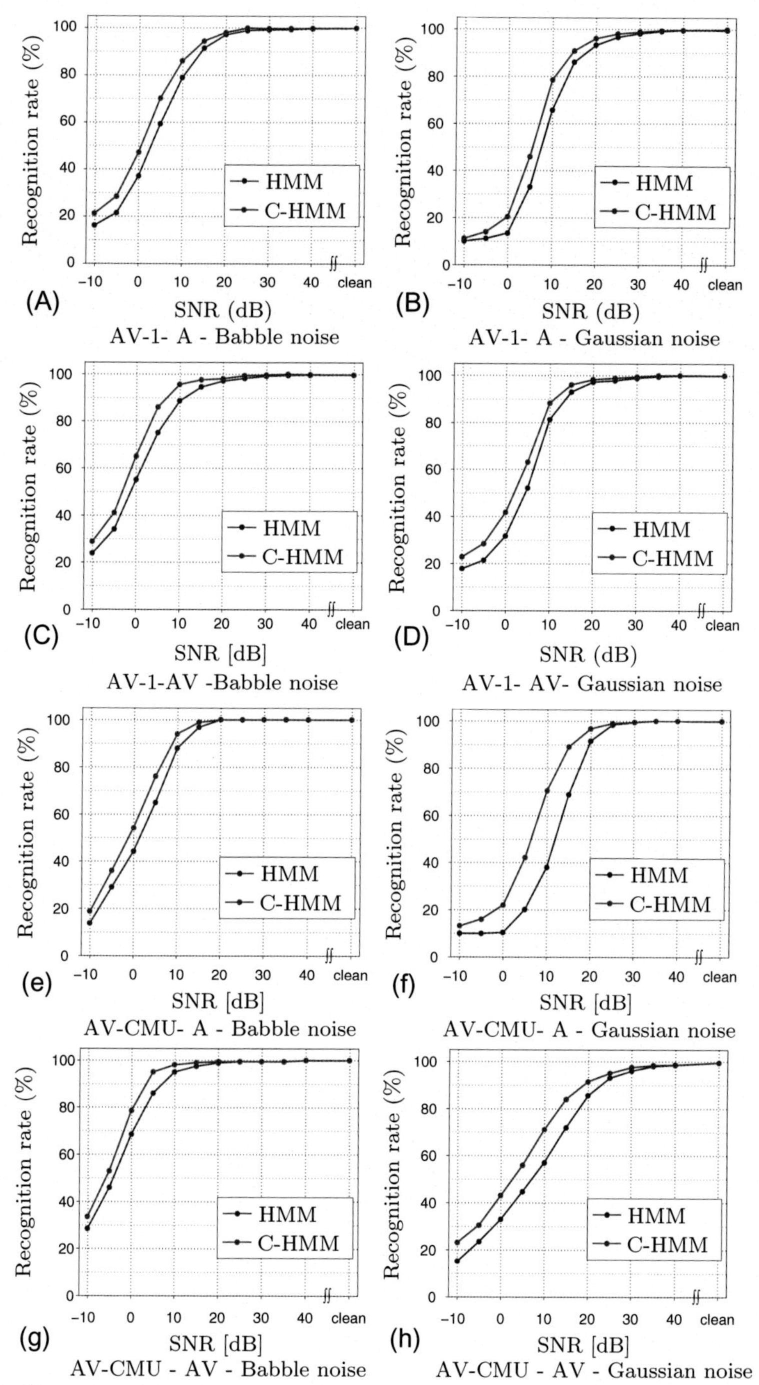

FIG. 4.9 Recognition based on HMM classifier and the proposed cascade of classifiers C-HMM. Recognition rates obtained over the AV-CMU and AV-UNR databases, using only acoustic (A) information and audio-visual (AV) information, for the cases of considering babble noise (first column) and Gaussian noise (second column).

TABLE 4.1 Lip-Reading Based on HMM Classifier

Database	Visual Features	HMM	C-HMM
AV-UNR	Mouth shape parameters [50]	70.16%	74.04%
AV-CMU	Lip contour model [10]	57.79%	61.82%
AVLetters	Local spatiotemporal descriptors [18]	57.30%	61.25%

Recognition rates based on HMM classifier and the proposed cascade of classifiers C-HMM over AV-UNR, AV-CMU and AVLetters.

In this case, there are four parameters to adjust, namely, the number of trees in the ensemble (N) and the size of the random subset of features used at each splitting node in the tree (α), for both, the RFs and the CRFs. It is understood that if the number of trees in the ensemble is large enough (usually 500 or more), the particular value employed does not significantly affect the performance of the RF classifier. In this chapter, the value of N was set to 2000. Finally, only two tuning parameters remain in the proposed system, namely, the size of the random subset of features used at each splitting node in the tree (α) for the RFs, and the CRFs, which were optimized via exhaustive search using values in the range from 2 to 10, looking for the combination giving the best recognition results.

The recognition rates obtained at different SNRs over the AV-CMU and AV-UNR databases, using audio-only information and fused audio-visual information are depicted in Fig. 4.10, for RF and the proposed cascade of classifiers (C-RF). It is clear that, for both databases, both types of inputs and both kinds of acoustic noise, the proposed scheme (C-RF) performs better than the one based on RF. This is more obvious at low- and middle range SNRs. In this case, the maximum improvement achieved by resorting to the proposed cascade of classifiers is almost 16%, for AV-UNR database using audio-only information and Gaussian noise with SNR = 0 dB. The results for the lip-reading scenario for AV-UNR, AV-CMU and AVLetters databases, are shown in Table 4.2. As it can be observed, the use of the proposed cascade of classifiers scheme (C-RF) improves the recognition rates for all databases. In this case, a significant improvement of 7% for AVLetters database is achieved.

4.6.3 Support Vector Machine

In this section, the method proposed in Section 4.4 for discriminative models is applied to SVMs [55]. Both stages of the cascade of classifiers proposed are implemented with SVM, based on Gaussian kernels. The complementary model based on SVM will be hereafter referred to as complementary support vector machines (CSVMs). There are four tuning parameters in this case, namely,

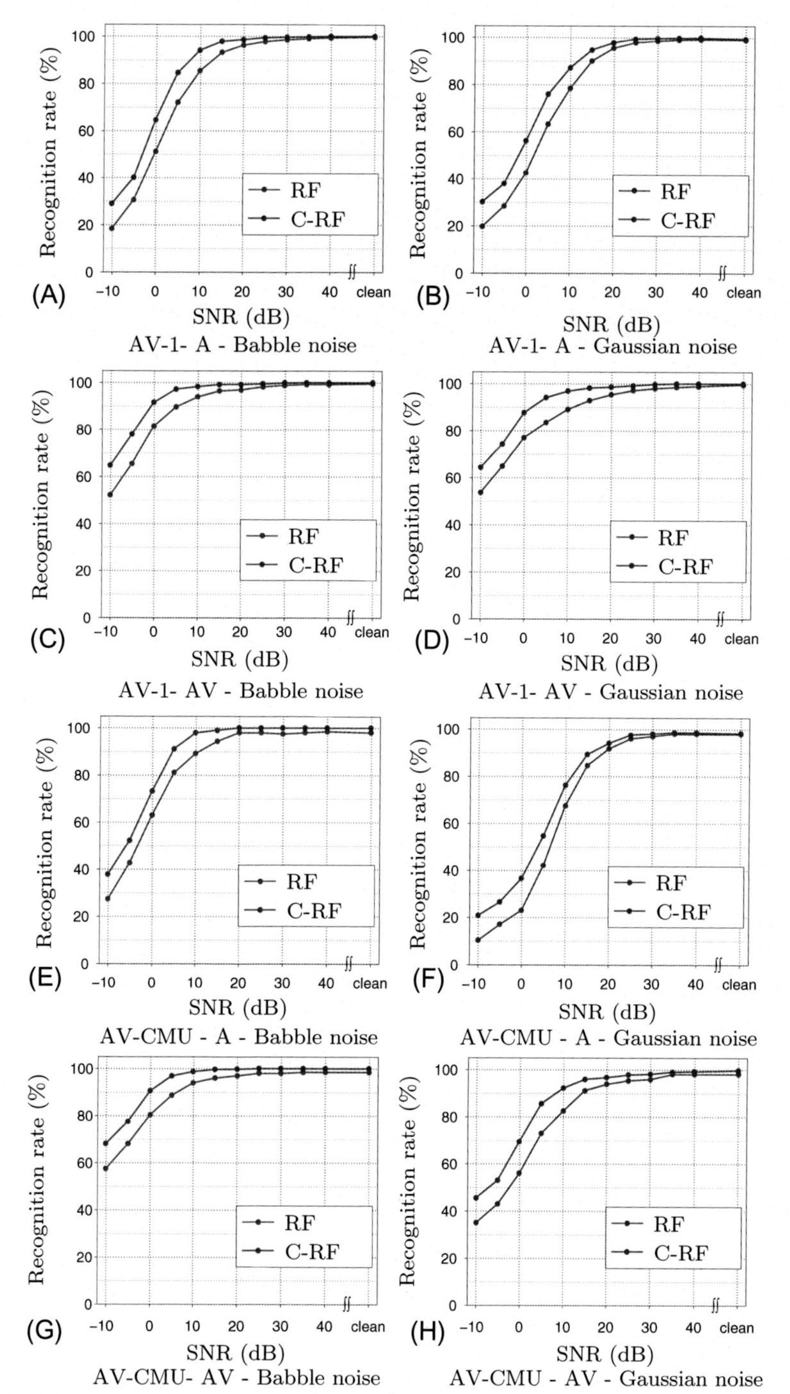

FIG. 4.10 Recognition based on RF classifier and the proposed cascade of classifiers C-RF. Recognition rates obtained over the AV-CMU and AV-UNR databases, using only acoustic (A) information and audio-visual (AV) information, for the cases of considering babble noise (first column) and Gaussian noise (second column).

TABLE 4.2 Lip-Reading Based on RF Classifier

Database	Visual Features	RF	C-RF
AV-UNR	Mouth shape parameters [50]	88.67%	91.97%
AV-CMU	Lip contour model [10]	71.65%	75.05%
AVLetters	Local spatiotemporal descriptors [18]	65.38%	72.34%

Recognition rates based on RF classifier and the proposed cascade of classifiers C-RF over AV-UNR, AV-CMU and AVLetters.

the cost C and the σ value of the Gaussian kernel, for both classifiers SVM and CSVM, which were optimized via exhaustive search in a two-stage procedure. First, a rough search was carried out, varying the C and σ values in decade steps ($[\ldots, 10^{-2}, 10^{-1}, 1, 10^{1}, 10^{2}, \ldots]$). Finally, a finer search was carried out, using smaller steps for the C and σ values, where the best performance of the recognition system was achieved in the C/σ parameters space at the first stage.

The recognition rates obtained at different SNRs over the AV-CMU and AV-UNR databases, using audio-only information and fused audio-visual information are depicted in Fig. 4.11, for SVM and the proposed cascade of classifiers (C-SVM). It is clear that, for both databases, both types of inputs and both kinds of acoustic noise, the proposed scheme (C-SVM) performs better than the one based on SVM. This is more evident at low- and middle range SNRs. In this case, the proposed scheme performs better for the case of babble noise in comparison with the case of Gaussian noise. The maximum improvement achieved by using the proposed cascade of classifiers is almost 13%, for the case of the AV-CMU database using fused audio-visual information and Gaussian noise with SNR = 0 dB. The results for the lip-reading scenario for AV-UNR, AV-CMU and AVLetters databases, are shown in Table 4.3. As it can be observed, the use of the proposed cascade of classifiers scheme (C-SVM) improves the recognition rates for all databases. In this case, a significant improvement of approximately 7% for AVLetters database is achieved.

4.6.4 AdaBoost

In this section, the method proposed in Section 4.4 for discriminative models is applied to boosting-based classifiers. Both stages of the cascade of classifiers proposed are implemented using the adaptive boosting algorithm (ADA) [17]. The complementary model based on ADA will be hereafter referred to as complementary AdaBoost (CADA). In this case, there are four tuning parameters, namely, the depth of each tree in the ensemble (d) and the number of iterations of the boosting algorithm (N), for both, the ADA and the CADA classifiers, which were optimized via exhaustive search. Several experiments

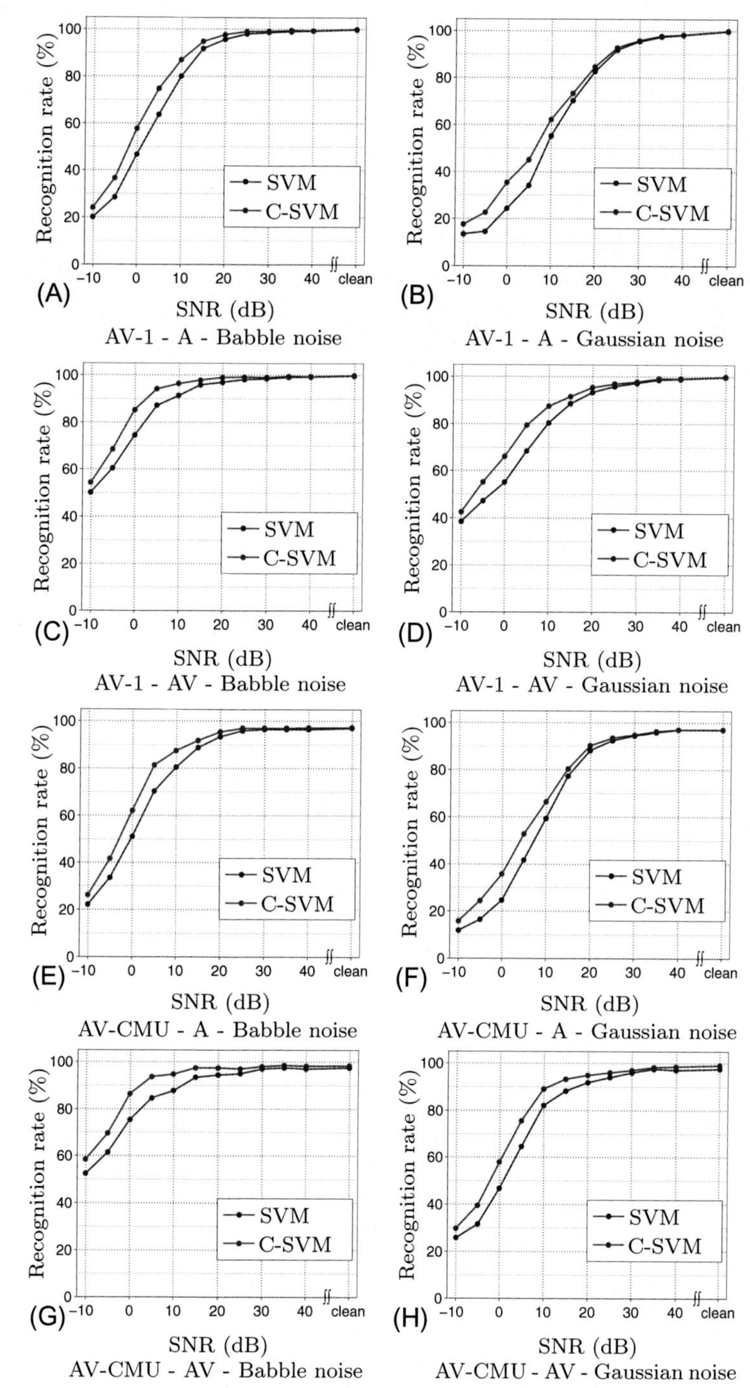

FIG. 4.11 Recognition based on SVM classifier and the proposed cascade of classifiers C-SVM. Recognition rates obtained over the AV-CMU and AV-UNR databases, using only acoustic (A) information and audio-visual (AV) information, for babble noise (first column) and Gaussian noise (second column).

TABLE 4.3 Lip-Reading Based on SVM Classifier

Database	Visual Features	SVM	C-SVM
AV-UNR	Mouth shape parameters [50]	83.75%	88.24%
AV-CMU	Lip contour model [10]	67.53%	73.15%
AVLetters	Local spatiotemporal descriptors [18]	63.08%	69.97%

Recognition rates based on SVM classifier and the proposed cascade of classifiers C-SVM over AV-UNR, AV-CMU, and AVLetters.

were carried out using values of d in the range from 2 to 20, and considering $N=[100, 500, 1000, 2000, 5000]$, looking for the combination that gives the best recognition results.

The recognition rates obtained at different SNRs over the AV-CMU and AV-UNR databases, using audio-only information and fused audio-visual information are depicted in Fig. 4.12, for the case of ADA and the proposed cascade of classifiers (C-ADA). It is clear that, for both databases, both types of inputs and both kinds of acoustic noise, the proposed scheme (C-ADA) performs better than the one based on ADA. This is obvious at middle SNRs. The maximum improvement by resorting to the proposed cascade of classifiers is almost 13%, for the case of AV-CMU database using fused audio-visual information and Gaussian noise with SNR = 15 dB. The results for the lip-reading scenario for AV-UNR, AV-CMU and AVLetters databases, are shown in Table 4.4. As it can be observed, the use of the proposed cascade of classifiers scheme (C-ADA) improves the recognition rates for all databases. In this case, a significant improvement of approximately 6.3% for AVLetters database is achieved.

4.6.5 Analysis and Comparison

As the results demonstrate, the proposed cascade of classifiers, based on complementary models, achieved recognition rate improvements compared with the traditional model, regardless of the experimental scenario setup (model/database/input information/kind of noise). This is apparent at low- and middle-range SNRs. Comparing the results obtained for each of the models used (C-HMM, C-RF, C-SVM and C-ADA), it can be observed that in general the improvements obtained were greater for the case of C-HMM and C-RF, reaching significant improvements in some cases (33% for C-HMM). Apart from the magnitude of the improvements obtained in the recognition rates by resorting to the proposed scheme, it is interesting to note that in none of the experiments, the results obtained by the cascade of classifiers were worse than that obtained with the traditional model. The cascade strategy proposed in this chapter could be thought

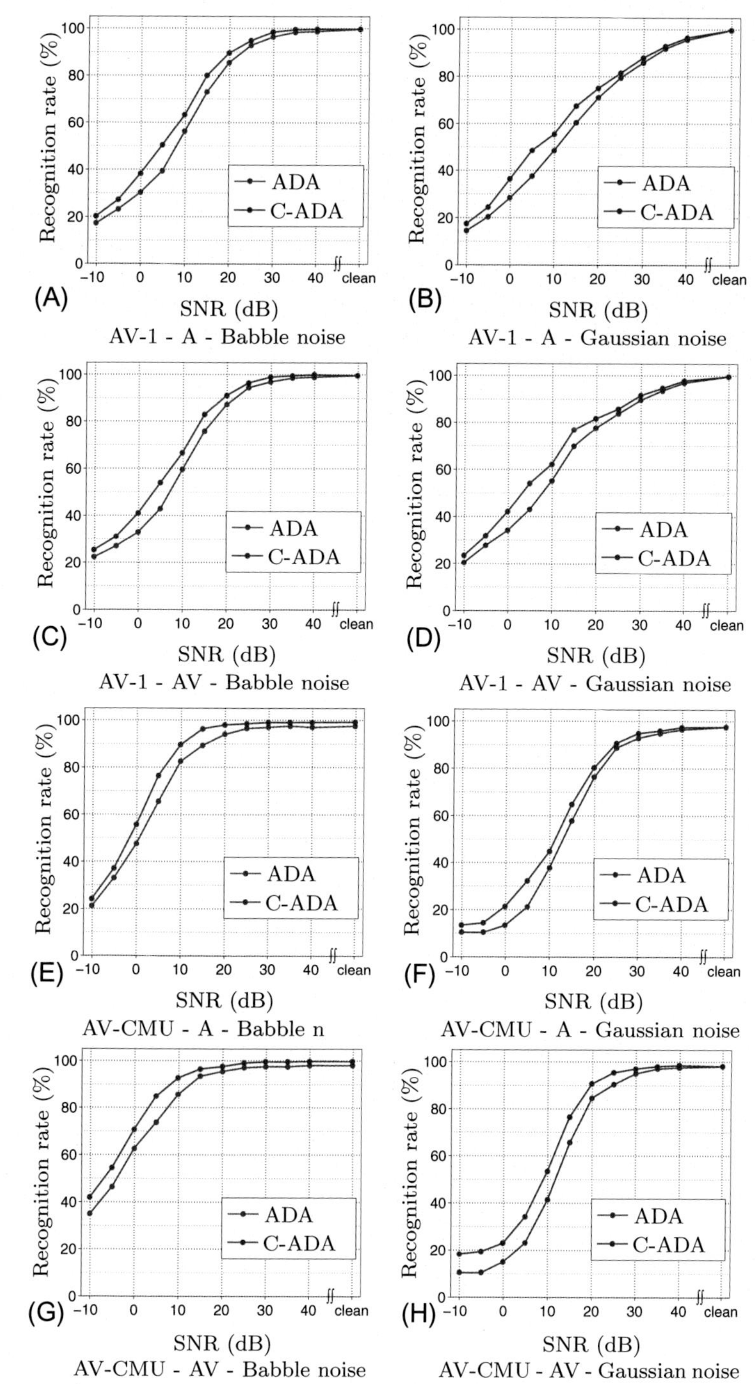

FIG. 4.12 Recognition based on ADA classifier and the proposed cascade of classifiers C-ADA. Recognition rates obtained over the AV-CMU and AV-UNR databases, using only acoustic (A) information and audio-visual (AV) information, for considering babble noise (first column) and Gaussian noise (second column).

TABLE 4.4 Lip-Reading Based on ADA Classifier

Database	Visual Features	ADA	C-ADA
AV-UNR	Mouth shape parameters [50]	85.63%	90.65%
AV-CMU	Lip contour model [10]	67.53%	72.24%
AVLetters	Local spatiotemporal descriptors [18]	54.23%	60.57%

Recognition rates based on an ADA classifier and the proposed cascade of classifiers C-ADA over AV-UNR, AV-CMU and AVLetters.

of as a block that is added to the output of the traditional model. That is, the proposed strategy could be applied to obtain recognition rate improvements on an existing model simply by training the complementary models, without any modification on the traditional models.

In Table 4.5, the recognition rates obtained for the lip-reading scenario over the AVLetters database by the proposed C-HMM, C-RF, C-SVM and C-ADA schemes are presented. Also, the results obtained with others methods proposed in the literature are shown. As can be observed, the proposed cascade of classifiers based on complementary models performs satisfactorily and better than the other approaches, independently of the model being used. In Table 4.6, the recognition rates obtained for the lip-reading scenario over the AV-CMU database by the proposed C-HMM, C-RF, C-SVM and C-ADA systems, are

TABLE 4.5 Lip-Reading Over AVLetters Database

Classifier	Visual Features	Accuracy
C-HMM	Local spatiotemporal descriptors [18]	61.25%
C-RF	**Local spatiotemporal descriptors** [18]	**72.34%**
C-SVM	Local spatiotemporal descriptors [18]	69.57%
C-ADA	Local spatiotemporal descriptors [18]	60.57%
HMM [45]	Multiscale spatial analysis [45]	44.60%
HMM [18]	Local spatiotemporal descriptors [18]	57.30%
SVM [18]	Local spatiotemporal descriptors [18]	58.85%
Deep Autoencoder [24]	Video-only deep autoencoder [24]	64.40%
RTMRBM [22]	Mouth region PCA [22]	64.63%

Recognition rates obtained with the proposed cascade of classifiers for the case of HMM (C-HMM), RF (C-RF), SVM (C-SVM), ADA (C-ADA), and also by others methods in the literature.

TABLE 4.6 Lip-Reading Over AV-CMU Database

Classifier	Visual Features	Accuracy
C-HMM	Lip contour model [10]	61.82%
C-RF	**Lip contour model [10]**	**75.05%**
C-SVM	Lip contour model [10]	73.15%
C-ADA	Lip contour model [10]	72.24%
HMM [10]	Lip contour model [10]	61.17%

Recognition rates obtained with the proposed cascade of classifiers for the case of HMM (C-HMM), RF (C-RF), SVM (C-SVM), ADA (C-ADA), and also by other method in the literature.

depicted. The results obtained with other methods proposed in the literature are also shown. Again, the proposed cascade of classifiers based on complementary models performs satisfactorily and better than the other approaches, independently of the model being used.

4.7 Conclusions

This chapter proposes a novel scheme for audio-visual speech classification tasks, based on the combination of traditional and complementary models in a cascade structure, to improve recognition rates. This proposed scheme can be employed either for different generative models or discriminative models. Also, it can handle different kinds of input information, viz., audio, visual or audio-visual information, indistinctly. The concept of complementary models is introduced, which is based in a novel training procedure. The main idea is to detect the absence of a class by redefining classes, and to this end, for each particular word in the vocabulary, a new class is defined using all the instances of the words in the vocabulary except the one corresponding to the one that is being represented. Four different models, viz., HMMs, RF, SVMs and adaptive boosting, were employed to analyze the efficiency of the proposed cascade of classifiers. The performance of the proposed speech classification scheme was evaluated at different conditions, considering only audio information, only video information (lip-reading), and fused audio-visual information. These evaluations were carried out over three different audio-visual databases, two of them public ones and the remaining one compiled by the authors of this chapter. The robustness of the proposed scheme against noisy conditions in the acoustic channel is also evaluated. Experimental results show that a good performance is achieved with the proposed system over the three databases and for the different kinds of input information being considered. Recognition rate improvements are achieved by means of the proposed cascade of classifiers in all the scenarios and for the four models used. In addition, the proposed method performs better than other reported methods in the literature over the same two public databases. The proposed strategy could be applied to obtain recognition rates improvements on an existing model simply by training the complementary models, without any modification on the traditional models.

References

[1] H. McGurk, J. MacDonald, Hearing lips and seeing voices, Nature 264 (5588) (1976) 746–748.

[2] P. Wagner, Z. Malisz, S. Kopp, Gesture and speech in interaction: An overview, Speech Comm. 57 (2014) 209–232.

[3] A. Jaimes, N. Sebe, Multimodal human-computer interaction: a survey, Comput. Vis. Image Underst. 108 (1–2) (2007) 116–134.

[4] A.K. Katsaggelos, S. Bahaadini, R. Molina, Audiovisual fusion: challenges and new approaches, Proc. IEEE 103 (9) (2015) 1635–1653.

[5] S. Shivappa, M. Trivedi, B. Rao, Audiovisual information fusion in human computer interfaces and intelligent environments: a survey, Proc. IEEE 98 (10) (2010) 1692–1715.

[6] R.K. Aggarwal, M. Dave, Acoustic modeling problem for automatic speech recognition system: advances and refinements (part ii), Int. J. Speech Technol. 14 (4) (2011) 309–320.

[7] R.K. Aggarwal, M. Dave, Acoustic modeling problem for automatic speech recognition system: Conventional methods (part i), Int. J. Speech Technol. 14 (4) (2011) 297–308.

[8] A.I. Amrous, M. Debyeche, A. Amrouche, Robust arabic speech recognition in noisy environments using prosodic features and formant, Int. J. Speech Technol. 14 (4) (2011) 351.

[9] A. Biswas, P.K. Sahu, M. Chandra, Multiple cameras audio visual speech recognition using active appearance model visual features in car environment, Int. J. Speech Technol. 19 (1) (2016) 159–171.

[10] B. Borgström, A. Alwan, A low-complexity parabolic lip contour model with speaker normalization for high-level feature extraction in noise-robust audiovisual speech recognition, IEEE Trans. Syst. Man Cybern. 38 (6) (2008) 1273–1280.

[11] V. Estellers, M. Gurban, J. Thiran, On dynamic stream weighting for audio-visual speech recognition, IEEE Trans. Audio Speech Lang. Process. 20 (4) (2012) 1145–1157.

[12] M. Miki, N. Kitaoka, C. Miyajima, T. Nishino, K. Takeda, Improvement of multimodal gesture and speech recognition performance using time intervals between gestures and accompanying speech, EURASIP J. Audio Speech Music Process. 2014 (1) (2014) 2.

[13] G. Papandreou, A. Katsamanis, V. Pitsikalis, P. Maragos, Adaptive multimodal fusion by uncertainty compensation with application to audiovisual speech recognition, IEEE Trans. Audio Speech Lang. Process. 17 (3) (2009) 423–435.

[14] N. Puviarasan, S. Palanivel, Lip reading of hearing impaired persons using HMM, Expert Syst. Appl. 38 (4) (2011) 4477–4481.

[15] G. Potamianos, C. Neti, G. Gravier, A. Garg, A.W. Senior, Recent advances in the automatic recognition of audio-visual speech, Proc. IEEE 91 (2003) 1306–1326.

[16] J. Shin, J. Lee, D. Kim, Real-time lip reading system for isolated Korean word recognition, Pattern Recogn. 44 (3) (2011) 559–571.

[17] R.E. Schapire, Y. Singer, Improved boosting algorithms using confidence-rated predictions, in: Machine Learning, 1999, pp. 80–91.

[18] G. Zhao, M. Barnard, M. Pietikäinen, Lipreading with local spatiotemporal descriptors, IEEE Trans. Multimedia 11 (7) (2009) 1254–1265.

[19] G. Potamianos, C. Neti, G. Iyengar, A.W. Senior, A. Verma, A cascade visual front end for speaker independent automatic speechreading, Int. J. Speech Technol. 4 (3) (2001) 193–208.
[20] L.D. Terissi, G. Sad, J.C. Gómez, M. Parodi, Audio-visual speech recognition scheme based on wavelets and random forests classification, in: Lecture Notes in Computer Science, vol. 9423, 2015, pp. 567–574.
[21] M.R. Amer, B. Siddiquie, S. Khan, A. Divakaran, H. Sawhney, Multimodal fusion using dynamic hybrid models, in: IEEE Winter Conference on Applications of Computer Vision, 2014, pp. 556–563.
[22] D. Hu, X. Li, X. Lu, Temporal multimodal learning in audiovisual speech recognition, in: 2016 IEEE Conference on Computer Vision and Pattern Recognition (CVPR), 2016, pp. 3574–3582.
[23] H. Meutzner, N. Ma, R. Nickel, C. Schymura, D. Kolossa, Improving audio-visual speech recognition using deep neural networks with dynamic stream reliability estimates, in: 2017 IEEE International Conference on Acoustics, Speech and Signal Processing (ICASSP), 2017, pp. 5320–5324.
[24] J. Ngiam, A. Khosla, M. Kim, J. Nam, H. Lee, A. Ng, Multimodal deep learning, in: Proceedings of the 28th International Conference on Machine Learning (ICML-11), 2011, pp. 689–696.
[25] K. Noda, Y. Yamaguchi, K. Nakadai, H.G. Okuno, T. Ogata, Audio-visual speech recognition using deep learning, Appl. Intell. 42 (4) (2015) 722–737.
[26] S. Petridis, M. Pantic, Deep complementary bottleneck features for visual speech recognition, in: 2016 IEEE International Conference on Acoustics, Speech and Signal Processing (ICASSP), 2016, pp. 2304–2308.
[27] S. Yin, C. Liu, Z. Zhang, Y. Lin, D. Wang, J. Tejedor, T.F. Zheng, Y. Li, Noisy training for deep neural networks in speech recognition, EURASIP J. Audio Speech Music Process. 2015 (1) (2015) 2.
[28] S. Ahmadi, S.M. Ahadi, B. Cranen, L. Boves, Sparse coding of the modulation spectrum for noise-robust automatic speech recognition, EURASIP J. Audio Speech Music Process. 2014 (1) (2014) 36.
[29] G. Monaci, P. Vandergheynst, F.T. Sommer, Learning bimodal structure in audio-visual data, IEEE Trans. Neural Netw. 20 (12) (2009) 1898–1910.
[30] P. Shen, S. Tamura, S. Hayamizu, Multistream sparse representation features for noise robust audio-visual speech recognition, Acoust. Sci. Technol. 35 (1) (2014) 17–27.
[31] J. Wright, Y. Ma, J. Mairal, G. Sapiro, T.S. Huang, S. Yan, Sparse representation for computer vision and pattern recognition, Proc. IEEE 98 (6) (2010) 1031–1044.
[32] J.A. Bilmes, K. Kirchhoff, Generalized rules for combination and joint training of classifiers, Pattern. Anal. Applic. 6 (3) (2003) 201–211.
[33] L. Deng, X. Li, Machine learning paradigms for speech recognition: An overview, IEEE Trans. Audio Speech Lang. Process. 21 (5) (2013) 1060–1089.
[34] J. Kittler, M. Hatef, R.P.W. Duin, J. Matas, On combining classifiers, IEEE Trans. Pattern Anal. Mach. Intell. 20 (3) (1998) 226–239.
[35] T.G. Dietterich, Ensemble methods in machine learning, in: Proceedings of the First International Workshop on Multiple Classifier Systems, MCS '00, Springer-Verlag, London, UK, 2000, pp. 1–15.
[36] M.J.F. Gales, D.Y. Kim, P.C. Woodland, H.Y. Chan, D. Mrva, R. Sinha, S.E. Tranter, Progress in the CU-HTK broadcast news transcription system, IEEE Trans. Audio Speech Lang. Process. 14 (5) (2006) 1513–1525.

[37] T. Hain, L. Burget, J. Dines, G. Garau, V. Wan, M. Karafi, J. Vepa, M. Lincoln, The AMI system for the transcription of speech in meetings, in: 2007 IEEE International Conference on Acoustics, Speech and Signal Processing (ICASSP), vol. 4, 2007, pp. 357–360.
[38] M. Hwang, W. Wang, X. Lei, J. Zheng, O. Cetin, G. Peng, Advances in mandarin broadcast speech recognition, in: 8th Annual Conference of the International Speech Communication Association, INTERSPEECH 2007, 2007, pp. 2613–2616.
[39] S. Stüker, C. Fügen, S. Burger, M. Wölfel, Cross-system adaptation and combination for continuous speech recognition: the influence of phoneme set and acoustic front-end, in: 9th International Conference on Spoken Language Processing (INTERSPEECH 2006-ICSLP), 2006, pp. 521–524.
[40] R.K. Aggarwal, M. Dave, Integration of multiple acoustic and language models for improved hindi speech recognition system, Int. J. Speech Technol. 15 (2) (2012) 165–180.
[41] C. Breslin, M. Gales, Directed decision trees for generating complementary systems, Speech Comm. 51 (3) (2009) 284–295.
[42] R. Hu, Y. Zhao, A bayesian approach for phonetic decision tree state tying in conversational speech recognition, in: 2007 IEEE International Conference on Acoustics, Speech and Signal Processing (ICASSP), vol. 4, 2007, pp. 661–664.
[43] A. Puurula, D. Van Compernolle, Dual stream speech recognition using articulatory syllable models, Int. J. Speech Technol. 13 (4) (2010) 219–230.
[44] J. Xue, Y. Zhao, Novel lookahead decision tree state tying for acoustic modeling, in: 2007 IEEE International Conference on Acoustics, Speech and Signal Processing (ICASSP), vol. 4, 2007, pp. 1133–1136.
[45] I. Matthews, T. Cootes, J.A. Bangham, S. Cox, R. Harvey, Extraction of visual features for lipreading, IEEE Trans. Pattern Anal. Mach. Intell. 24 (2002) 2002.
[46] F.J. Huang, T. Chen, Advanced Multimedia Processing Laboratory, Cornell University, Ithaca, NY, 1998. http://chenlab.ece.cornell.edu/projects/AudioVisualSpeechProcessing. Accessed July 2018.
[47] N. Krishnamurthy, J. Hansen, Babble noise: modeling, analysis, and applications, IEEE Trans. Audio Speech Lang. Process. 17 (7) (2009) 1394–1407.
[48] G.D. Sad, L.D. Terissi, J.C. Gómez, Complementary Gaussian mixture models for multimodal speech recognition, in: Third IAPR TC3 Workshop on Multimodal Pattern Recognition of Social Signals in Human-Computer Interaction (MPRSS 2014), Revised Selected Papers, Lecture Notes in Computer Science, vol. 8869, 2015, pp. 54–65.
[49] G.D. Sad, L.D. Terissi, J.C. Gómez, Class confusability reduction in audio-visual speech recognition using random forests, in: 22nd Iberoamerican Congress on Pattern Recognition (CIARP 2017), Lecture Notes in Computer Science, vol. 10657, 2018, pp. 584–592.
[50] L. Terissi, J. Gómez, 3D head pose and facial expression tracking using a single camera, J. Univ. Comput. Sci. 16 (6) (2010) 903–920.
[51] J. Ahlberg, Candide-3—An Updated Parameterized Face. Technical Report, Department of Electrical Engineering, Linkping University, Sweden, 2001.
[52] A. Varga, H.J.M. Steeneken, Assessment for automatic speech recognition II: NOISEX-92: A database and an experiment to study the effect of additive noise on speech recognition systems, Speech Comm. 12 (3) (1993) 247–251.
[53] L. Rabiner, A tutorial on hidden Markov models and selected applications in speech recognition, Proc. IEEE 77 (2) (1989) 257–286.
[54] L. Breiman, Random forests, Mach. Learn. 45 (1) (2001) 5–32.
[55] C. Cortes, V. Vapnik, Support-vector networks, Mach. Learn. 20 (3) (1995) 273–297.

Chapter 5

A Deep Dive Into Deep Learning Techniques for Solving Spoken Language Identification Problems

Himanish Shekhar Das and Pinki Roy
Department of Computer Science and Engineering, National Institute of Technology Silchar, Silchar, India

5.1 Introduction

Automatic language identification (LID) is a challenging research area in the domain of speech signal processing. It is the identification of a language from a random spoken utterance. The process of spoken LID is a front end for many applications such as multilingual conversational systems, spoken language translation, multilingual speech recognition, spoken document retrieval, human machine interaction through speech, etc. automatic speech recognition (ASR) systems are generally language dependent. Once a spoken LID system correctly identifies a language, an ASR system can take over to process the speech.

In his paper, Muthusamy [1], explained one critical use of automatic LID. It was about distress calls made to 911 operators. Since there are many U.S. citizens from different ethnic groups, it was found that in times of crisis, people used their own native language when calling for help. It is therefore necessary to understand what language the person is using before forwarding the call to an interpreter who can understand the message. AT&T introduced Language Line Services, which employed interpreters who could handle 140 languages. However, Muthusamy found that when a person who spoke Tamil, (a language spoken in India and Sri Lanka) called Language Line Services, there was a 3-min delay before the language was identified, and a Tamil interpreter was brought online. In such crucial circumstances, LID can play a crucial role.

Deep learning is currently used in many lines of. This is an era dominated by artificial intelligence (AI), specifically deep learning techniques such as in bioinformatics [2], forecasting the future of the mobile number portability [3], and

Intelligent Speech Signal Processing. https://doi.org/10.1016/B978-0-12-818130-0.00005-2

advancement in multimedia content [4]. In Spoken LID too many deep learning techniques have been used with significant success. Prominent among the deep learning techniques are feed-forward deep neural network (FF-DNN) commonly referred to as multilayer perceptron (MLP), convolutional neural network (CNN), long short term memory-recurrent neural network (LSTM-RNN), etc. These contemporary types of deep neural network techniques have overshadowed conventional methods such as Gaussian mixture model (GMM) and hidden Markov model (HMM). These techniques showed significant improvement in recognition performance over various parameters. It is the ability of deep neural networks techniques to perform complex correlation among speech signal features, which enhances their performance over traditional approaches.

5.2 Spoken Language Identification

While a human identifying a spoken language is commonplace, mimicking the process for machines is not. Let us first try to understand how a human identifies a language.

Humans are trained in one or more languages soon as they start learning. It is easy to identify the language(s) they learned if a person knows the phonemes, syllables, words, and sentence structure. The person get the cues about the language from multiple prompts, viz. phonetic, phonotactic, prosodic, and lexical. It is also important to note that even if a language that is similar in phonetic features to a language people know, they can make an educated guess about the language. In most cases, such a language belongs to the same family of languages. For example, if a person knows Bengali, languages like Assamese, Odiya, and Hindi (from Indo-Aryan language family), it will not very difficult for the person to identify even with very little hearing (small training set). However, it will not be that easy for the person to understand languages like Tamil, Telegu, and Kannada (from Dravidian language family), because these languages belong to another family of languages. Identifying other languages belonging to other families will require a higher degree of training. Then again, if somehow one can correlate some phones of the unknown language to a known language(s), the person can make out syllables and map them according to the known language. This is also relevant in case of written languages. How often do we see, (mostly in social media) that one can use English to construct Bengali words? Even if the person on the other end has very basic knowledge of English, one can interpret those words to Bengali and have an understanding. This is possible because phones of Bengali can be expressed in terms of English phones. In addition, we humans are excellent at comprehending multilingual speeches. Often we use a word from other languages in our speech, but it does not create a problem for a listener to identify the primary language of the speaker as well as tag those foreign words and separately identify them, too (only if one has previous knowledge about those foreign words). It can be concluded that we humans use acoustic models as well as language models apart from other sources of information to identify a language.

It is not that simple for a machine though. It is widely known in the speech processing community that making a machine proficient in all languages will evidently make the machine an excellent language identifier. The flip side is that it will incur a lot of expenses in terms of time, money, and labor. This is why different strategies have been formulated to identify a language by a machine.

All the spoken LID systems can be broadly classified into two groups: explicit and implicit LID systems. The systems, which require segmented and labeled speech corpus, are known as explicit LID systems. The systems that do not require phone-recognizers (thus no need for segmented and labeled speech data) are known as implicit LID systems. In other words, implicit LID systems require only the raw speech data along with the true identity of the language spoken.

5.3 Cues for Spoken Language Identification

As discussed earlier, there must be some unique feature(s), which will separate one language from another. Researchers like Muthusamy et al. [1]; Zissman and Berkling [5]; Li et al. [6]; Lee [7]; in their respective papers, almost unanimously categorized the cues into the following groups:

- *Acoustic Phonetics:* Whatever we say, at its smallest possible structure can be termed as a phone. According to Coxhead [8], a phone can be defined as a "unit sound" of a language. It is a "unit" sound because the whole of the phone must be substituted to make a different word. Phones can be again (for convenience) divided into monophones, diphones, and triphones among others. A monophone refers to a single phone, a diphone is an adjacent pair of phones and a triphone is simply a group of three phones. Acoustic phonetics is driven by the belief that a set of phones are different in different languages, and even if languages have identical phones, the frequency of occurrence will vary. The phones that comprise the word "celebrate" might be [s eh 1 ax bcl b r ey q]. Acoustic phonetic features are extracted by several methods. Some popular methods are linear predictive coding (LPC), linear prediction cepstral coefficients (LPCC), perceptual linear prediction coefficients (PLP), mel-frequency cepstral coefficient (MFCC), filter bank (FB), and shifted delta coefficients (SDC).

LPC is based on the source-filter model of speech signal. One can think of LPC as a coding method; a way of encoding information in a speech signal into a smaller space for transmission over a restricted channel. LPC encodes a signal by finding a set of weights on earlier signal values that can predict the next signal value. The LPC calculates a power spectrum of the signal. For a medium or a low bit rate coder, LPC is most widely used. While we pass the speech signal from speech analysis filter to remove the redundancy in signal, a residual error is generated as an output. It can be quantized by a smaller number of bits compared to the original signal. A parametric model is computed based on the least mean square error theory; this technique is known as linear prediction (LP). By this method, the speech signal is approximated as a linear combination of its previous samples. In this technique, the acquired LPC coefficients describe the formants. The frequencies at which the resonant peak occur are called formant frequencies. Thus with this method, the locations of the formants in a speech signal are estimated by computing the linear predictive coefficients over a sliding window and finding the peaks in the spectrum of the resulting LP filter.

Another important extension of the LPC is the LPCC, a short-term cepstral representation. The cepstral analysis is used to decorrelate dependences among variables of the acoustic features.

PLP models human speech based on the concept of psychophysics of hearing. PLP discards irrelevant information of speech and thus improves the speech recognition rate. PLP is identical to LPC except that its spectral characteristics have been transformed to match characteristics of the human auditory system. PLP approximates three main perceptual aspects namely: the critical-band resolution curves, the equal-loudness curve, and the intensity-loudness, power-law relation, which are known as cubic-root.

MFCC is the most used spectral feature extraction method. MFCCs are based on frequency domain using the mel scale, which is based on the human ear scale. MFCC is a representation of the real cepstral of a windowed short-time signal derived from the fast Fourier transform (FFT) of that signal. The difference from the real cepstral is that a nonlinear frequency is used. The MFCC coefficients are represented in a number of frames centered at equally spaced times, during a constant sampling period.

The human ear perceives sound in a nonlinear fashion. Filterbank analysis provides a straightforward route to obtaining the desired nonlinear frequency resolutions. An array of band-pass filters are used to separate the input signal into multiple components.

The SDCs, which are computed from the deltas of MFCCs, can capture a wide range of dynamics of an utterance. Therefore they showed significant improvement in performance of spoken LID.

- *Phonotactics:* Phones are a unit sound, but they do not exist as individual entities in speech. Every utterance is composed of multiple phones together in various combinations. Phonotactics describes the rules of combination of phones. It is believed that for every language, the rules of the combination of phones are different. For instance, the phone cluster /sr/ is very common in Tamil, but not in English.
    1. *Prosodics:* Prosodics are speech features such as stress, tone, or word juncture that accompany or are added over constants and vowels; these features are not limited to single sounds but often extend over syllables, words, or phrases. It is believed that prosody is useful for distinguishing

between broad language classes. Mandarin and Vietnamese have very different intonation characteristics than English.

2. *Vocabulary and Syntax:* Vocabulary is the collection of words, specific to a language. There will be words in every language that are unique and can directly point to the language. Even if there are common words between languages, their syntactic way to form a sentence will not be the same.

5.4 Stages in Spoken Language Identification

The process of spoken LID can be divided into two major blocks, feature extraction, and classification. However, there can be many other operations that can be done in speech signal preprocessing like voice activity detection and noise removal, feature extraction and classification are the most important stages of the activity.

In feature extraction (FE) stage cues which will help in discrimination between languages are extracted as features (Fig. 5.1). These features are then used in the classification stage to obtain a score, which helps in identifying the spoken language. Currently, there is a trend of end-to-end spoken LID systems. In end-to-end systems, there is no distinction between FE and classification stages; both of them are merged into the systems (Fig. 5.2).

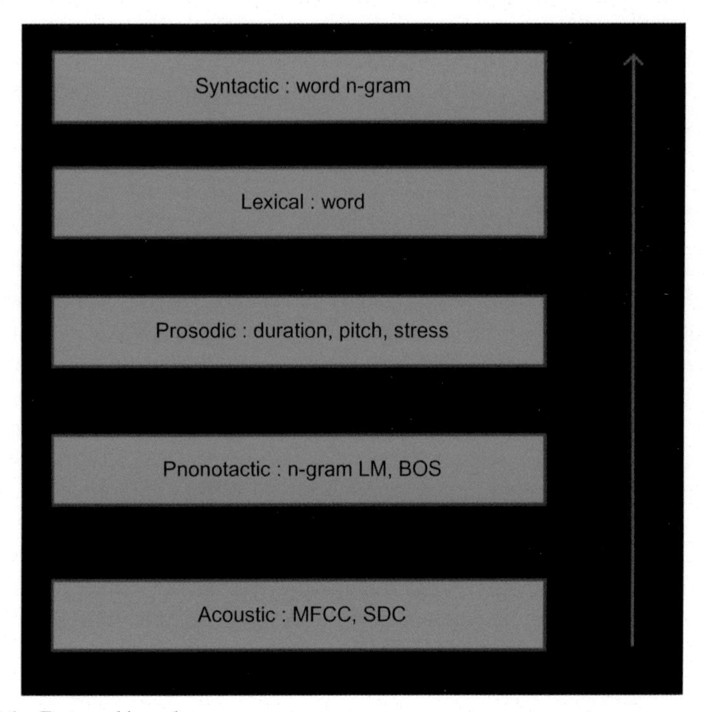

FIG. 5.1 Features hierarchy.

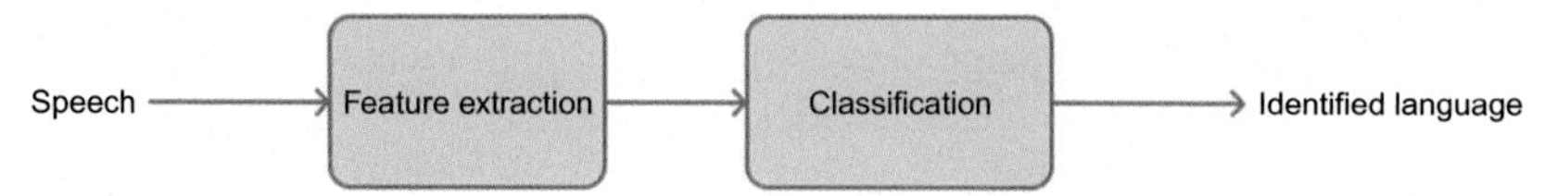

FIG. 5.2 LID process.

5.5 Deep Learning

In machine learning (ML), knowledge is acquired from data representations and not from specific formulae and algorithms. When learning is done, it is not in a single layer of computation, rather it spans across two or more layers. They type of learning is known as deep learning (DL). In the present scenario, deep learning and deep neural network are almost synonymous. If people feel interested in many facets of deep learning, they can read the Stanford University UFLDL tutorial [9], *Deep Learning*, by Goodfellow et al. [10], or *Neural Networks and Deep Learning* by Nielsen [11].

Since deep learning is inspired by biological neural network and human is still the best intelligence when it comes to identify a person in a picture or melody in a song or whether it is to an extent safe to jump over a ditch. It is for this reason that deep learning is thought to be suitable over traditional machine learning algorithms. Since most of traditional machine learning concepts use a domain expert approach, how the machine will learn is determined by the logic handcrafted by humans. The algorithms are extremely important in decision making. Whereas in deep learning, very little formulation is done and learning is instead, dependent on data. Availability of more data generally results in better accuracy. Deep learning does a remarkable job when the problem is complex and data availability is high.

5.6 Artificial and Deep Neural Network

An artificial neural network (ANN) is an aspect of AI that is focused on emulating the learning approach that humans use to gain certain types of knowledge. Like biological neurons, which are present in the brain, ANN also contains a number of artificial neurons, and uses them to identify and store information. ANN consists of input and output layers, as well as (in most cases) one or more hidden layer(s) consisting of units that transform the input into something that the output layer can use (Fig. 5.3).

The human brain contains approximately 86 billion neurons [12]. These neurons are connected to each other like a mesh. Stimuli from the external environment or inputs from sensory organs are accepted by entities known as dendrites. These inputs create electric impulses, which quickly travel through the neural network. A neuron can then send messages to other neuron(s) to handle the issue or does not send it forward. This is known as activation of the neuron. A neuron is connected to thousands of other neurons by axons.

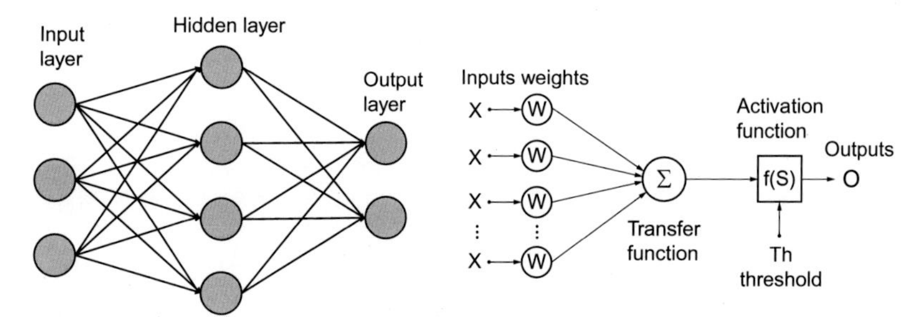

FIG. 5.3 A typical ANN and a typical artificial neuron.

ANNs are composed of multiple artificial neuron nodes, which imitate biological neurons of the human brain. Unlike biological neurons, there is only one type of link that connects one neuron to others. The neurons take input data and simple operations are performed on those data. The result of these operations is passed to other neurons. Whether the result will be passed, is determined by the activation function. The activation function plays an important role for both feature extraction and classification.

In the biological neural network, the size of dendrites varies in accordance with the importance of inputs. In ANN, this is achieved by using weights and biases. Earlier experimental work checked the performance of ANN with SVM and found that ANN performs better than SVM [13].

ANN can be of many types. Prominent among them are feed-forward neural network, radial basis function neural network, Kohonen self-organizing neural network, recurrent neural network, convolutional neural network and modular neural network.

A deep neural network (DNN) can be considered as stacked neural networks, i.e., networks composed of several layers.

- *FF-DNN:* FF-DNN, also known as multilayer perceptrons (MLP), are as the name suggests DNNs where there is more than one hidden layer and the network moves in only forward direction (no loopback). These neural networks are good for both classification and prediction. For spoken LID, we use the classification approach. When the FF-DNN is used as a classifier, the input and output nodes will match the input features and output classes.

The most important concepts in a FF-DNN are weights, biases, nonlinear activation and backpropagation. Let us try to understand them. Below is a sample FF-DNN (Fig. 5.4).

The input layer has four elements I-1, I-2, I-3, and I-4. They define an input, I = {I-1, I-2, I-3, I-4}. What we need is to find one or more patterns from the entities of the input, so that those patterns can be used to classify one output from the other. In order to do that, we devise a number of hidden units with activation function. Nonlinear activation functions like Sigmoid, tanh, ReLU are used to form a pattern of active hidden units.

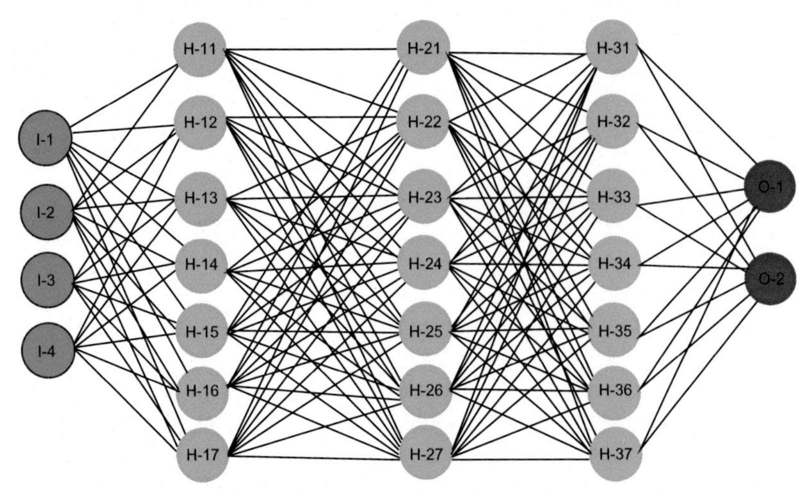

FIG. 5.4 Feed-forward deep neural network with three hidden layers.

Let us assume that, H-13, H-14, H-17, H-22, H-23, H-26, H-32, H-33, H-36, and H-37 are needed for active output O-1. To achieve this, we need to adjust the weights and biases such that, these hidden nodes are activated by the activation function. Initially, weights and biases are randomly initialized. Then, we train the network with tens of thousands of inputs. We use backpropagation of error to adjust weights and biases, so that suitable values of them will activate the hidden neurons.

The set of weights and biases, which will be used to uniquely identify a particular output, are called feature set or kernel.

It is believed that when the pattern used for discrimination is so complex that traditional statistical and numerical approaches do not work, FF-DNN is the key to solution.

- *Radial Basis Function Neural Network:* The radial basis function neural networks (RBF-NN) are inspired by biological neural systems, in which neurons are organized hierarchically in various pathways for signal processing. They are tuned to respond selectively to different features/characteristics of the stimuli within their respective fields. RBF-NN is structurally the same as MLP. RBF-NN has an input layer, a hidden layer, and an output layer. RBF-NN is strictly limited with exactly one hidden layer. The hidden units provide a set of functions that constitute an arbitrary basis for the input patterns. Hidden units are known as radial centers and the hidden layers are feature vectors. There are different radial functions like, Gaussian radial function, thin plate spline, quadratic, inverse quadratic, etc. The most popular radial function is Gaussian activation function. RBF is used in pattern classification and regression.

- *Kohonen Self-Organizing Neural Network:* Self-organizing neural networks can cluster groups of similar patterns into a single set. They assume a topological structure among their cluster units effectively mapping weights to input data. Kohonen self-organizing map neural network is one of the basic types of self-organizing maps. The ability to self organize provides new possibilities-adaptation to formerly unknown input data. It seems to be the most natural way of learning, which is used in our brains, where no patterns are defined. These patterns take shape during the learning process. Self-organizing networks can be either supervised or unsupervised. Unsupervised learning is a means of modifying the weights of a neural network without specifying the desired output for any input patterns. The advantage is that it allows the network to find its own solution, making it more efficient with pattern association. The disadvantage is that other programs or users have to figure out how to interpret the output. The functioning of a self-organizing neural network is divided into three stages: construction, learning, and identification.
- *Recurrent Neural Network:* FF-DNN only moves in the forward direction. No feedback loop is used. But there are many elements like speech that works in sequence. You cannot make out the meaning of a sentence without all the words, and even if you have them, you need them to be sequenced to find the correct meaning. Recurrent networks, unlike FF-DNN, take as their input not just the current input, but also state perceived previously in time. The idea behind RNNs is to make use of sequential information. In theory RNNs can make use of information in arbitrarily long sequences, but in practice they are limited to looking back only a few steps (Fig. 5.5).

The left side shows a representative form of a folded RNN, which means the network is available for n iterations in time, while the right side shows RNN states in times $t-1$, t, and $t+1$. The right side is also known as unrolled RNN and represents a full network. Let us understand how RNN works. x_t is the input at time step t. s_t is the hidden state at time step t. s_t contains a previous

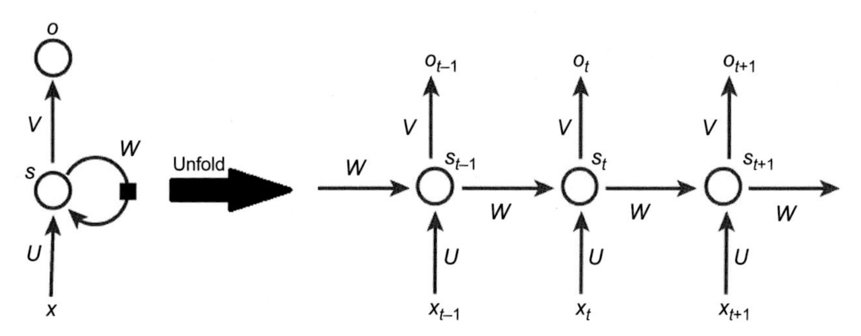

FIG. 5.5 Recurrent neural network.

hidden state, and is calculated taking consideration of the previous hidden state along with the input at the current step. $s_t = f(Ux_t + Ws_{t-1})$. The function f is usually a nonlinearity such as tanh or ReLU. $s - 1$, which is required to calculate the first hidden state, and is typically initialized to all zeroes. o_t is the output at step t. unlike a traditional deep neural network, which uses different parameters at each layer, an RNN shares the same parameters, U, V, and W; across all steps. This reflects that we are performing the same task at each step, just with different inputs. This greatly reduces the total number of parameters needed to be learned. Training an RNN is similar to training a traditional neural network. A backpropagation algorithm is used with a little twist. This is called backpropagation through time (BPTT). One of the most popular implementations of RNN is long short-term memory (LSTM) which solves the vanishing/exploding gradient problem.

- *CNN:* CNN works a little differently from FF-DNN and RNN where a neuron is supposed to do an activation. In case of a CNN, a neuron is a result of multiple convolution operations before getting activated for feature extraction or classification. A CNN has multiple stages of operation, viz., convolution, pooling, nonlinearity. These stages can be repeated several times to make a deep CNN. In the convolution stage, convolution operation is performed, which can be simply stated as the element-wise multiplication between the filter elements and the input. The input is represented as a two-dimensional matrix for that purpose. A filter or kernel is also a two-dimensional matrix, but of a smaller size. There may be a weight filter and a bias filter. The elements of the weight filter are multiplied with input nodes and the elements of the bias filter are added. To reduce the dimension of the output, a pooling operation is done. The most widely used pooling operation is max-pooling. Then a nonlinearity like ReLU or Sigmoid or tanh is used as an activation function. Finally, when the layers are done, a fully connected (FC) layer is used for classification (Fig. 5.6).
- *Modular Neural Network:* A modular neural network is a combination of independent neural networks. Each independent neural network serves as a module to the system as a whole. The task at hand is divided into subtasks, which are optimized by individual modules. The reason behind is

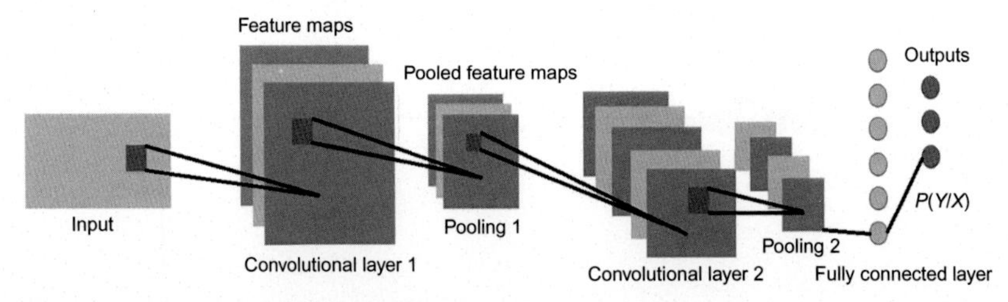

FIG. 5.6 Convolutional neural network.

that every neural network has its strength, and if it is required to improve the performance of the overall system, all its components need to be optimized in their operation.

5.7 Comparison of Spoken LID System Implementations with Deep Learning Techniques

- *FF-DNN for Spoken LID:* Perhaps the most cited FF-DNN with an acoustic-phonetic feature is that of Lopez-Moreno et al. [14]. They used Google 5M LID corpus and NIST LRE 2009 corpora. The target test utterance was 3 s. ReLU was the activation function. Two and eight hidden layers were used in two different test sets. Each hidden layer contained 2560 units. One extra output node was provided for out-of-set languages. Thirty-nine PLP ($13+\Delta+\Delta\Delta$) features were used in frame level. The input layer was fed with 21 frames formed by stacking the current processed frame and its ± 10 left-right context. Training was done using asynchronous stochastic gradient descent (SGD) within the DistBelief framework. The learning rate was 0.001, and minibatch size was 200 samples. The output scores were computed at an utterance level by respectively averaging the log of the softmax output of all its frames. C_{avg} (average cost) and EER (equal error rate) was used as performance metrics. It was claimed that the best performance was achieved by the DNN systems, where the eight hidden layers of DNN architecture yielded up to a ~70% relative improvement in C_{avg} terms with respect to the best reference system with Google 5M LID corpus. A relative improvement of ~43% in EER was obtained with eight-hidden layers of DNN, trained with 200 h of NIST LRE 2009 corpora. It was concluded with a heavy amount of training data (>20 h) per language is a key for the DNN to outperform i-vector system.

Richardson et al. [15] proposed another FF-DNN for language identification. PLP features were chosen as the feature. Samples of speech were converted into 20 ms fragments. A context of ±5–10 frames around the current input frame was used. DNN posteriors and DNN bottleneck features were used. For the experiments, a subset of six languages of the NIST 2009 LRE corpora was used. Twenty hours of training per language was done resulting in 120 h total training. Samples of speech at 3, 10, and 30 s were selected for experimentation. For a direct approach, DNN was trained using the training partition of the data set and consisted of 819 input nodes (stack of 21 frames of 13 Gaussianised PLP with their first- and second-order derivatives). Two hidden layers with 2560 nodes per layer and six output nodes were selected. The activation function was chosen to be Sigmoid. The learning rate was kept at 0.2. Degradation in performance was observed relative to the baseline system at the 30 and 10 s durations,. There was a slight gain at 3 s duration for two-layer DNN. BNF together with GMM posterior provided the best performance. It was observed that increasing the amount of training data and the number of parameters both up to a factor of three did not significantly improve performance. Appending delta features to the PLP features gave a small but significant performance gain. It was mentioned that without Guassianisation on the input PLP features, the LID performance degraded by well over a factor of two. Fusion did not always show a better identification rate. The best domain score fusion showed a performance gain of almost 20% relative to the BNF/GMM system alone, while the best out of domain score fusion gave a relative gain of only about 9%.

Herry et al. [16] proposed a different method based on language pair discrimination using MLP as classifiers of acoustic features. OGI MLTS Corpus of 11 languages was used. The focus was on 3 s signals. Several MLPs were used where each of them discriminated a couple of languages. The theory behind it is that less output choices improves accuracy, so a language is paired with one of the rest of the 10 languages to make one detection phase. Finally, all the results were merged to conclude the identified language. The front end was RASTA processing to generate acoustic vectors of dimension 24 of relevant spectrum representation. RASTA processing on 32 ms signal with an overlap of 50% was used. The MLPs used sigmoidal activation functions, 50 hidden cells and two outputs. The learning algorithm used was the stochastic back propagation algorithm. The obtained results highlighted scores ranging from a 75.1% to an 82% detection rate.

- *CNN for Spoken LID:* Montavon [17] proposed a CNN for spoken LID. He used VoxForge and RadioStream dataset. Each sample corresponded to a speech signal of 5 s. One shallow CNN-TDNN and one three-layer deep CNN-TDNN were used for experiment. 30 mel-frequency filterbank between 0 and 5 kHz was used. The whole 5 s sample was used as an input. The classifier was trained with SGD. A confusion matrix was used for performance measurement. It was claimed that deep architecture was 5%–10% more accurate than the shallow counterpart. It was also concluded that the overall accuracy of a spoken LID system can vary considerably depending on the selected subset of languages to be identified. It was surmised that deep learning, compared to techniques based on hand-coded features, is still a mystery. It was suggested that accuracy can be improved by collecting more samples and extending the time dependence in order to learn higher level language features.

Lei et al. [18] proposed a CNN implementation in noisy conditions. In one of the approaches, the state posterior counts from the CNN are used directly for language identification. The approach was inspired by both the phoneme posteriogram and PRLM approaches. They used Defense Advanced Research Projects Agency (DARPA) Robust Automatic Transcription of Speech program (RATS) data. Speech samples taken at 3, 10, 30 and 120 s are used for the experiment. Three models, posterior extraction using CNN, CNN/i-vector and CNN posterior, were used. Forty filterbanks with a context of 15 frames were deployed, the height of the convolutional filter was kept at eight. Two hundred convolutional filters were used to model the data in more detail. The output vectors of the different filters are concatenated into a long vector that is then input to a traditional DNN. This DNN usually includes five to seven hidden layers. CNN trained for ASR is used to extract the posterior for every frame. Then the posteriors from the CNN are used to estimate the zeroth and first order statistics for i-vector model training. This is an acoustic way CNN/i-vector system. A CNN posterior system focus is on modeling the sequence of phonetic units given by a phone recognizer. This approach only uses zeroth order statistics to estimate the low-dimensional vector. Five target languages and one that was from the set language were taken for the study. 200×8 convolutional filters of 40 log mel filterbank coefficients with a context of 7 frames from each side of the center frame were chosen as input. The subsequent DNN included five hidden layers with 1200 nodes per hidden layer, and the output layer with 3353 nodes of representation of the senones. The neural network back-ends with a single hidden layer of 200 nodes were used for all presented system. All input vectors to the NN back-end were size 400. The output layer was a size 6. To optimize the performance on all durations, the original dataset was separated into 8 and 30 s segments with a 50% overlap. The performance was evaluated on three measurements, including EER, for all target languages. The miss rate where the false alarm rate was equal to 1% and the C_{avg}. fusion with linear logistic regression and twofold cross-validation were also done. Relative improvements between 23% and 50% on average EER with respect to a state-of-the-art UBM/i-vector system across different duration conditions. Fusion of both approaches gives to a 20% relative performance gain over the best individual system.

Ganapathy et al. [19] also supported using CNN for robust spoken LID. Though, they confined the use of CNN as a feature extractor and not as a classifier. The CNN models used were trained on noisy data provided under the RATS program for Arabic Levantine (ALV) and Farsi (FAS) KWS. Three hundred hours of data for each language was used for training. The CNNs were trained on 32 dimensional log-mel spectra augmented with Δ and $\Delta\Delta$s. The log-mel spectra were extracted by first applying the mel scale integrators on power spectral estimates in short analysis windows (25 ms) of the signal followed by the log transform. Each frame of speech was also appended temporally with a context of 11 frames. The CNNs used two convolutional layers with 512 hidden nodes. The first convolutional layer was processed with 9×9 filters while the second convolutional layer was processed with 4×3 filters. The nonlinear outputs from the second convolutional layer were then input to a fully connected DNN. Three hidden layers with 2048 units, followed by a bottleneck layer with 25 activations were deployed. They concluded that fusing bottleneck features along with acoustic features provided good results. The CNN BN features provided robust representations which were useful for the spoken LID tasks. The BN features provided about 21% relative improvement in the evaluation set and about 25% in the development set, as claimed by the authors.

- *RNN for Spoken LID:* Gonzalez-Dominguez et al. [20] proposed an LSTM RNN implementation for spoken LID. NIST LRE 2009 corpora was used for training and testing. Eight representative languages were selected with up to 200 h of audio available. Focus was given to 3 s utterances. The LSTM RNN architecture used contained 512 memory cells. The input to the network was 39-dimensional PLP features that were calculated at a given time step with no stacking of acoustic frames. The total number of parameters N, ignoring the biases, was $N = n_i \times n_c \times 4 + n_c \times n_c \times 4 + n_c \times n_o + n_c \times 3$, where n_c is the number of memory cells, n_i is the number of input units and n_o is the number of output units. The LSTM RNN model was trained using asynchronous stochastic gradient descent (ASGD), and the truncated backpropagation through time (BPTT) learning algorithm within a distributed training framework. Activations are forward propagated for a fixed-step time of 20 over a sub sequence of an input utterance; the cross entropy gradients are computed for this subsequence and backpropagated to its start. For better randomization of gradients in ASGD and stability of training, the training utterances were split into random chunks of length between 2.5 and 3 s. The same language identification was set sparsely for a chunk, one in every five frames for the experiments. The errors were only calculated for the frames for which a target language identification were set. One hundred machines were used for distributed training, and in each machine four concurrent threads processed a batch of four subsequences. An exponentially decaying learning rate of le-04 was used. For scoring, an utterance level score was computed for each target language by averaging the log of the softmax outputs for that target language of all the frames in an utterance. In order to assess the performance, two different metrics were used, C_{avg} and ERR. Two major findings were concluded. It was observed that LSTM RNN architecture provided better performance than FF-DNN with four hidden layers. The fact is particularly interesting as the proposed LSTM RNN contained 20 times fewer parameters than four hidden layers FF-DNN. Also, the C_{avg} values indicated that the scores produced by the LSTM RNN model were better calibrated than those produced by FF-DNN or i-vector systems. Both neural network approaches (FF-DNN and LSTM-RNN) surpass the i-vector system performance by ~47% and ~52% in EER and C_{avg}, respectively.
- *Hybrid NN Models for Spoken LID:* Every ANN has its strengths. When applied in a particular task, one NN provides better performance than others do. Keeping that in consideration, researchers tried hybrid NN models for the spoken LID.

Jin et al. [21] proposed an end-to-end DNN-CNN classification for spoken LID. The authors introduced two end-to-end DNN-CNN neural network variants, which utilized high-order LID-senone statistics. Both systems combine the benefits of both the high-order Baum-Welch statistics calculation of i-vector systems with the natural discriminating attributes of neural networks. In LID-net2, high-order statistics were obtained through an O2P method borrowed from fine-grained visual recognition, whereas in LID-bilinear-net, the statistics were obtained using the outer product operation from two different layers and pooled to obtain an utterance representation. Experiments were conducted on a full 23 languages of NIST LRE 2009 corpora, and performances were compared to the state-of-the-art DBF/i-vector systems. First, a six-layer DNN (48×21-2048-2048-50-2048-2048-3020) with an internal bottleneck layer was trained. Then parameters were transferred from the first three layers to DNN layers 1–3 of LID-net, which was trained. Finally, all layer parameters below the SPP layer were transferred to LID-net2 and LID-bilinear-net and the two networks were trained separately. In LID-net2, the high-order statistics of LID-senones were obtained through second-order pooling (O2P) of posteriors pooled from the CNN convolution layer prior to the FC mapping network. In LID-bilinear-net, the network utilized posteriors pooled from two different CNN layers instead of one. Each network was trained and tested independently for 30, 10 and 3 s data. Each network had six convolutional layers. The feature maps from CNN layers 1–5 had 512 channels and the feature maps after layer six were evaluated with between 64 and 512 channels. Each convolutional layer was followed by a batch normalization layer and first and second order LID-senone statistics were evaluated. It was observed that both systems outperformed the baseline system for 3 and 10 s utterances in terms of EER. Only 30 s utterances were a slightly behind in terms of performance.

Ranjan et al. [22] proposed two-stage training and transfer of learning in their research. They used the dataset from the NIST 2015 LRE i-vector machine learning challenge. The training set contained 300 i-vectors per language corresponding to each of the 50 in set target languages along with a test set and a development set. The speech utterances corresponding to the training set i-vectors were chosen so that their durations would exhibit a log-normal distribution with a mean duration of 35.15 s. In the two-step DNN training strategy, the initial DNN was trained using only in set labeled training data. The initial DNN was then used to estimate out-of-set labels from the development data. Next, a second DNN is trained for LID with both in set and estimated out-of-set labels. A FF-DNN with sigmoid activation function was used in first training. A new FF-DNN is trained using the i-vectors in the second stage. The mini-batch SGD algorithm was used with a mini-batch size of 256. The DNN had two hidden layers with 1024 units each. The input layer had 400 nodes corresponding to the 400-dimensional i-vectors. A dropout factor of 0.2 was used. The output layer had 50 nodes corresponding to all in set languages. A cost value of 26.82 was reported for 1024 hidden layers.

In their paper, Ferrer et al. [23] compared three approaches using combinations of DNN with other systems to identify the spoken language. The first two approaches used DNN as a feature extractor, while in the third case, DNN is used as classifier. In the first approach, zeroth-order statistics for i-vector extraction were estimated using a DNN, and the method was named DNN/iv modeling approach. The second approach, named DNN/post, used a DNN output layer to create features for language recognition. The posteriors for each senone at each frame were processed to generate a single vector per utterance, which is then modeled with standard back-end techniques. A third approach was to train a DNN with a bottleneck architecture, and then use the outputs of the bottleneck layer as features within an i-vector framework. NIST LRE 2009 and RATS SLR were used as corpus. The authors claimed improvements between 40% and 70% relative to a state of art GMM i-vector system on test durations from 3 to 120 s.

Bartz et al. [24] used a hybrid convolutional recurrent neural network (CRNN) that operates on spectrogram images of audio snippets. In their work, they tried to utilize the power of CNNs to capture spatial information, and the power of the RNNs to capture information through a sequence of time steps for identifying the language from a given audio snippet. They created their own dataset. Data was collected for six different languages. The network architecture consisted of two parts. The first part was the convolutional extractor that took a spectrogram image representation of the audio file as input. This feature extractor convolved the input image in several steps and produced a feature map with a height of one. The feature map was then sliced along the x-axis, and each slice was used as a time step for the subsequent BLSTM network. The network used five convolutional layers, each layer followed by the ReLU activation function, batch normalization and 2×2 max pooling with a stride of 2. The BLSTM consisted of two single LSTMs with 256 outputs units each. Finally, both outputs were concatenated to a vector of 512 dimensions and fed into a fully connected layer with 4/6 output units serving as the classifier. The authors claimed an accuracy of 92% and an F1 score of 0.92 (Table 5.1).

5.8 Discussion

A total of 11 highly appreciated and cited Research Papers on automatic LID with deep learning are discussed above. This leads to an understanding of a few very useful criterions for a successful deep learning-based sutomatic LID. First, in order to attract more researchers to implement deep learning-based language identification system, we need a standardized, free and openly available data corpus. NIST is currently being used widely, but its availability is a factor that hinders more research in this area [15, 16, 23]. OGI MLTS is of age and a new consortium must be formed to make available a free and open speech corpora. Second, PLP and Filterbanks are still used widely, even though MFCC and its acceleration coefficients are considered as de facto standard. Third, almost every author is looking for a fusion of approaches, and fused approaches are providing better accuracies. Ferrer et al. [23] refers to a fused i-vector and

TABLE 5.1 Summary of the Works

Author Reference	Feature Extraction Technique	Classification Technique	Corpora Used	No. of Languages	Recognition Accuracy
13	39 PLP (13 + Δ + Δ Δ) Stack of 21 frames	FF-DNN with 2–8 hidden layers	Google 5M and NIST LRE 2009	Google 5M – 25 languages +9 Dialects NIST LRE 2009 – 8 languages	EER – 9.58 (average) for DNN_8_200h
14	39 PLP (13 + Δ + Δ Δ) Stack of 21 frames	FF-DNN with 2 hidden layers	NIST LRE 2009	6 languages	C_{avg} – 2.56 (for 30 s) 4.24 (for 10 s) 10.3 (for 3 s)
15	RASTA	MLP with 50 hidden cells	OGI MLTS	11 languages	Average Competitive Rate – 77%
16	End-to-end		VoxForge and RadioStream	3 languages	VoxForge – 91.2% (known speakers) 80.1% (new speakers) RadioStream – 87.7% (known radios) 83.5% (new radios)
17	40 Filterbanks	i-vector, posterior	RATS data	5 languages	EER – CNN/posterior – 14.08 (3 s) CNN/i-vector – 13.60 (3 s)

Continued

TABLE 5.1 Summary of the Works—cont'd

Author Reference	Feature Extraction Technique	Classification Technique	Corpora Used	No. of Languages	Recognition Accuracy
18	Multiple features including CNN-BF	GMM/i-vector	RATS data	5 languages	EER – BN-ALV – 15.3 (3 s) BN-FAS – 15.0 (3 s)
19	39 PLP	LSTM RNN	NIST LRE 2009	8 languages	EER –8.35 (average) C_{avg} – 0.0944
20	End-to-end		NIST LRE 2009	23 languages	EER – LID-net-256 – 7.57 (3 s) LID-bilinear-net-512-relu – 6.86 (3 s)
21	300 i-vector	DNN	NIST 2015 LRE i-vector machine learning challenge	50 languages	Cost (progressive subnet) DNN-2-1024 – 26.82
22	Mixed	Mixed	NIST LRE 2009 and RATS data	4 languages	DNN-eng/post – 14.18 (3 s)
23	End-to-end		Own dataset	4 languages	Accuracy – 92%

DNN BN system, which gives the best accuracy values in all 3, 10 and 30 s segments. Bartz et al. [24] also used a fusion of four systems to get the best C_{avg} values. A similar approach is observed in other papers, too. The reason is that all approaches have their strengths and weaknesses, and no single one can claim to be the best approach. The fusion of approaches is thus the only contender for the best approach. Fourth, its the 3 and 10 s segments that are critical for applications. Most of the work is done to find the suitable approach, which can result in best 3 s accuracy value. Finally, end-to-end systems are making headway and may be the future of automatic LID.

5.9 Conclusion

It is important to understand the role of spoken LID in speech technology. It is neither a complete ASR, nor does it involve the intricacies of NLP systems. Though it needs to be accurate, at the same time, it should also be fast and lightweight. Today, the spoken LID scenario can be divided into two branches, end-to-end and hybrid NN systems. Both have their advantages. In end-to-end, both feature extraction and classification are done with the same system, reducing the processing time and resource. But hybrid NNs tend to perform better. There has to be a compromise between accuracy and performance. Even though PPRLM and i-vector systems are still state-of-the-art, DNNs are catching on fast, and some results show that they even surpassed the other systems. Yet there is the question of how will the systems work when placed in a real-time environment. We have seen different sets of speech data corpus; NIST LRE is predominant, but even they do not cover all the languages of the world. People still need to create their own dataset for training and testing. An enormous repository of speech data is an absolute necessary before the systems can be used for field trials. There has been many parameters for accuracy but C_{avg} and EER are almost a benchmark now. Among the trending technologies, hybrid end-to-end DNNs seem to be the solution for the future. Since speech is a collection of sequences and not stationary, we may see variants of RNN and other sequence-training NNs in the future. We, the human beings, are mostly trained like a PPRLM model and that is why deep models that can emulate PPRLM, is something to look for in the future.

References

[1] Y.K. Muthusamy, E. Barnard, R.A. Cole, Automatic language identification: a review/tutorial, IEEE Signal Process. Mag. 11 (4) (1994) 33–41.

[2] K. Lan, D.T. Wang, S. Fong, L.S. Liu, K.K. Wong, N. Dey, A survey of data mining and deep learning in bioinformatics, J. Med. Syst. 42 (8) (2018).

[3] S. Hu, M. Liu, S. Fong, W. Song, N. Dey, R. Wong, Forecasting China future MNP by deep learning, in: Behavior Engineering and Applications, Springer, 2018, pp. 169–210.

[4] Dey, N., Ashour, A. S., & Nguyen, G. N., Deep Learning for Multimedia Content Analysis, Mining Multimedia Documents, CRC Press. ISBN 9781138031722.

[5] M.A. Zissman, K.M. Berkling, Automatic language identification, Speech Comm. 35 (1–2) (2001) 115–124.
[6] H. Li, B. Ma, K.A. Lee, Spoken language recognition: from fundamentals to practice, Proc. IEEE 101 (5) (2013) 1136–1159.
[7] C.-H. Lee, Principles of spoken language recognition, in: Springer Handbook of Speech Processing, Springer, Berlin, Heidelberg, 2008, pp. 785–796.
[8] P Coxhead, Natural Language Processing & Applications, In lecture Notes, NLP Resources.
[9] UFLDL Tutorial, Stanford University, http://deeplearning.stanford.edu/wiki/index.php/UFLDL_Tutorial.
[10] I. Goodfellow, Y. Bengio, A. Courville, Deep Learning, MIT Press, 2016.
[11] M. Nielsen, Neural Networks and Deep Learning, http://neuralnetworksanddeeplearning.com.
[12] Wikipedia, https://en.wikipedia.org/wiki/List_of_animals_by_number_of_neurons.
[13] P. Roy, P.K. Das, S.K. Gupta, Comparison of SVMs and NNs approach for automatic identification of Indian languages, in: 1st International Science & Technology Congress, IEMCONGRESS-2014, August 28–31, Science City, Kolkata, India, 2014, ISBN 978-93-5107-248-5, pp. 39–44.
[14] I. Lopez-Moreno, J. Gonzalez-Dominguez, D. Martinez, O. Plchot, J. Gonzalez-Rodriguez, P.J. Moreno, On the use of deep feedforward neural networks for automatic language identification, Comput. Speech Lang. 40 (2016) 46–59.
[15] F. Richardson, D. Reynolds, N. Dehak, Deep neural network approaches to speaker and language recognition, IEEE Signal Process. Lett. 22 (10) (2015) 1671–1675.
[16] S. Herry, B. Gas, C. Sedogbo, J.-L. Zarader, Language detection by neural discrimination, in: Proc. ICSLP, 2004.
[17] G. Montavon, Deep learning for spoken language identification, in: NIPS Workshop on Deep Learning for Speech Recognition and Related Applications, 2009, pp. 1–4.
[18] Y. Lei, L. Ferrer, A. Lawson, M. McLaren, N. Scheffer, Application of convolutional neural networks to language identification in noisy conditions, in: Proc. Odyssey-14, Joensuu, Finland, vol. 41, 2014.
[19] S. Ganapathy, K. Han, S. Thomas, M. Omar, M. Van Segbroeck, S.S. Narayanan, Robust language identification using convolutional neural network features, in: *Fifteenth Annual Conference of the International Speech Communication Association*, 2014.
[20] J. Gonzalez-Dominguez, I. Lopez-Moreno, H. Sak, J. Gonzalez-Rodriguez, P.J. Moreno, Automatic language identification using long short-term memory recurrent neural networks, in: Fifteenth Annual Conference of the International Speech Communication Association, 2014.
[21] M. Jin, Y. Song, I. McLoughlin, End-to-end DNN-CNN Classification for Language Identification, in: Proceedings of the World Congress on Engineering, vol. 1, 2017.
[22] S. Ranjan, C. Yu, C. Zhang, F. Kelly, J.H.L. Hansen, Language recognition using deep neural networks with very limited training data, IEEE International Conference on Acoustics, Speech and Signal Processing (ICASSP) (2016) 5830–5834.
[23] L. Ferrer, Y. Lei, M. McLaren, N. Scheffer, Study of senone-based deep neural network approaches for spoken language recognition, IEEE/ACM Trans. Audio Speech Language Process. 24 (1) (2016) 105–116.
[24] C. Bartz, T. Herold, H. Yang, C. Meinel, Language identification using deep convolutional recurrent neural networks, in: International Conference on Neural Information Processing, Springer, Cham, 2017, pp. 880–889.

Chapter 6

Voice Activity Detection-Based Home Automation System for People With Special Needs

Dharm Singh Jat*, Anton Sokamato Limbo* and Charu Singh†
Namibia University of Science and Technology, Windhoek, Namibia, †Sat-Com (PTY) Ltd., Windhoek, Namibia

6.1 Introduction

Voice activity detection (VAD) is a technique in which the presence or absence of human speech is detected. The detection can be used to trigger a process. VAD has been applied in speech-controlled applications and devices like smartphones, which can be operated by using speech commands. According to the Namibia Statistics Agency (NSA) Disability Report [1], the 2011 census revealed that there are 98,413 people living with disability in Namibia. This report further states that about 64% of the people living with a disability use the radio as an ICT asset for communication, implying that these people are not utilizing other technologies available today that can automate processes in a home. Most voice-based home automation systems are based on remote control or smartphones to control the home, or they depend on commercial ASR (automatic speech recognition) application programing interface (API), which is intended for general use, therefore not specially designed for home automation commands [2].

Speech-controlled applications and devices that support human speech communication are becoming more and more popular. An example of this is in modern mobile devices like smartphones, which can be operated by using spoken commands for controlling home devices. In the automotive industry, speech-controlled applications are in hands-free telephony, and speech-controlled applications that enable the driver to interact with the car infotainment systems while driving without being distracted. VAD can also be applied to homes to automate day-to-day processes using voice commands. This approach can benefit a wide range of people, especially people who face challenges in doing everyday tasks like operating electronic appliances in a home. This can include people living

Intelligent Speech Signal Processing. https://doi.org/10.1016/B978-0-12-818130-0.00006-4

with physical disabilities or with special needs, and the elderly; anyone who would who would appreciate the convenience of automated processes in a home. People with special needs and the elderly, especially those who live alone suffer the difficulty to control home appliances, and often need another person to assist them in performing everyday tasks like switching on a light.

A VAD system in homes can work by receiving voice commands from the user and taking the appropriate action based on the received spoken command, without requiring the user to have physical interaction with the system.

A lot of work has been done in the areas of voice processing and home automation, some of which have employed innovative approaches using Internet of Things (IoTs) to automate processes. Price et al. [3] propose an architecture that uses deep neural networks for ASR and VAD with improved accuracy, programmability, and scalability. The architecture is designed to minimize off-chip memory bandwidth, which is the main driver of system power consumption.

A new approach for real-time fault detection in a microphone array was proposed for reliable voice-based, human-robot interaction by intercorrelation of features in voice activity detection for each microphone [4].

In a fighter aircraft communication system, a voice operated switch (VOS) can be used for the optimal use of resources in the aircraft. This system provides a hands-free environment for a fighter pilot to perform some crucial tasks, and it can be a better option compared with a push to talk function. A reliable, robust VAD scheme was proposed for VOS applications in aircraft, which are based on fuzzy logic, energy-based detectors, and artificial neural networks (ANN). The VOS applications were compared with real-time robust VAD mechanisms [5]. Further, the evaluation of the VAD algorithms was performed by a MATLAB simulation platform using recordings of actual fighter aircraft cockpit noise.

The study presented a speech recognition system in robots designed for distant speech, which is useful for operating in human-occupied, outdoor military warehouse. Further, this study introduced a voice activity detector system by using multiple microphones and channel selection method [6].

Thanu and Uthra [7] a software system was developed which has features of interacting with home appliances remotely using the Internet. This allows users to have flexible control mechanisms remotely through a secured Internet web connection. Users are able to control appliances while away from home, or monitor the condition of these appliances.

Folea et al. [8] developed a smart home architecture, which employed a universal remote that will enable the user to control the alarm, lighting, and temperature in all rooms of the house. Furthermore, the system will be able to monitor dust in the air, if windows and doors are closed.

Soumya et al. [9] used a Raspberry Pi to develop a home monitoring system. The Raspberry Pi has a few components connected to it, which include an ultrasonic sensor and light dependent resistor (LDR), light emitting diodes (LEDs). When a person enters the room, the ultrasonic sensor detects a human presence, and based on whether the room is bright or dark, which is decided by the LDR, the LED turns on.

Xu et al. [10] developed a VAD algorithm that is highly sensitive to background noise. The algorithm calculates permutation entropy (PE), and determines the presence or absence of speech, as well as distinguishing between voiced and unvoiced parts of speech. Experiments done under several noise scenarios demonstrated that the proposed method can obtain conspicuous improvements on false alarm rates while maintaining comparable speech detection rates when compared to the reference method.

In order to suppress noise and reverberation, a direction of arrival estimation and localization (DOAE) is essential in many applications including multiparty teleconferencing for steering and beamforming and automatic steering of video cameras. This will also improve speech intelligibility [11].

As more and more physical devices are being connected to computer networks, home automation can employ IoT while using VAD as a way to interact with IoT devices using voice commands. The systems mentioned above use different approaches to home automation and speech processing, but lack the integration of using speech processing in home automation.

This chapter describes a VAD-based home automation system to assist people with special needs or some elderly people to provide support to control home appliances. This chapter details the overall design of a VAD-based home automation system, which can be integrated with existing infrastructure of home appliances. The system determines the suitable VAD algorithm for people who need assistance, especially those who live alone.

6.2 Conceptual Design of the System

The system will be designed to integrate with existing home appliances while requiring little physical interaction with the user. Fig. 6.1 shows the conceptual design of the voice activity detection-based home automation system for people with special needs.

In the conceptual design of the system, the user gives a voice command to the system, which is captured by the microphone and sent to the module responsible for detecting voice and performing noise suppression on the received speech. The noise suppressed speech is transferred to another module, which is responsible for processing the speech by converting it to text. The text is then analyzed to determine to see if it contains known commands. If so, the system then triggers an automation process that is in line with the received spoken command. Fig. 6.2 shows the flowchart of the VAD-based home automation system, and explains the process of capturing speech signals, reducing background noise, and activating the devices.

For the system to be fully operational, a few additional components must be added to ensure that the system is able to make the decisions by interacting with speech. This includes having other sensors and actuators working concurrently with the system to ensure that all operations are automated. The implementation of the VAD-based system is explained in the next section.

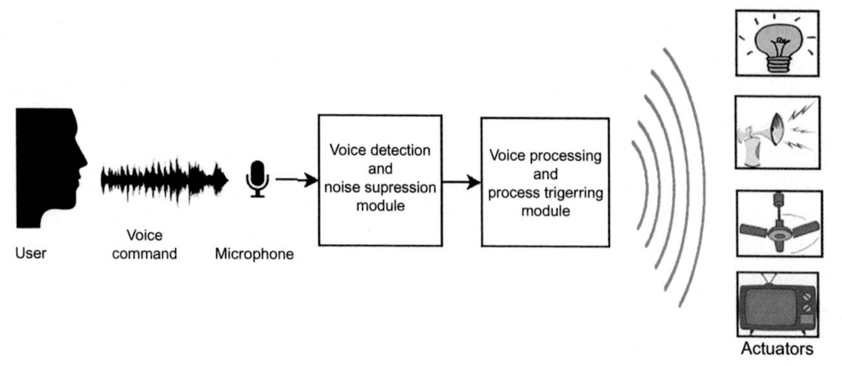

FIG. 6.1 Conceptual design of the VAD-based home automation system.

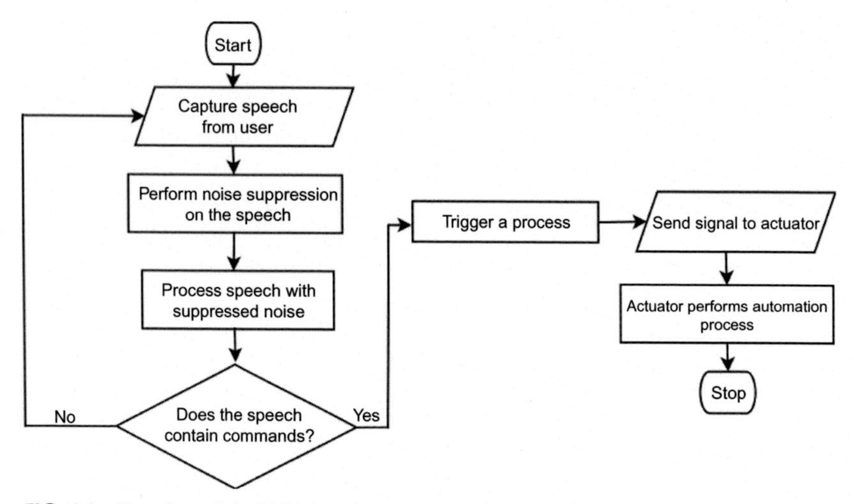

FIG. 6.2 Flowchart of the VAD-based home automation system.

6.3 System Implementation

The home automation system is implemented using a Raspberry Pi (RP) 3 Model B+, shown in Fig. 6.3.

The RP is the central point of the system capable of capturing the spoken commands and processing these commands by converting them into text using a Speech-to-Text engine. The converted text can then be used as the basis of the action that the system will take.

Specifications of a Raspberry Pi 3 include a 1.2 GHZ quad-core processor, 1 GB RAM, 4 USB ports and communication modules such as an Ethernet port, Wireless LAN, and Bluetooth. The most useful part of the Raspberry Pi, which will make it very ideal for an automation project, are the general purpose input output (GPIO) pins. They enable the user to connect the sensors and transmitters that will be communicating to the home appliances.

FIG. 6.3 Experimental setup for the VAD-based home automation system.

The system employs an LED infrared transmitter and infrared receiver, which is connected to the Raspberry Pi. The infrared receiver will be used to synchronize an appliance remote control with the automation system. The LED infrared transmitter will act like a remote control by emitting an infrared signal from the system to an appliance. Other hardware components required include a USB microphone, and because the Raspberry Pi has multiple communications modules, the system can be configured to control other appliances that do not have infrared by employing components like infrared controlled power plugs, light bulbs, and electric fans. In cases where devices cannot be operated using infrared, the system can control smart devices using Wireless LAN, or Bluetooth. This is discussed later in this chapter.

6.3.1 Speech Recognition

The CMU's fastest PocketSphinx speech recognition system is able to translate speech in real-time. PocketSphinx is an open source speech recognition system, and it is a library written in C language. It is the most accurate engine for real-time application, and therefore it is a good choice for home Automation live applications [12] [13]. The chapter describes an application of controlling a home device from a research and commercial perspective.

PocketSphinx captures speech in a waveform, then splits it at utterances, maybe a cough, an "uh," or a breath in a speech. It then tries to recognize what is being said in each utterance. To achieve this, PocketSphinx takes all the possible combinations of words and tries to match them with the audio. Then, it chooses the best matching combination.

PocketSphinx is also capable of performing noise suppression. For this reason, the automation system employs PocketSphinx as both a noise-suppression model and a Speech-to-Text engine. This is done by using keyword spotting and by using an acoustic and language model. By using keyword spotting as a wake word, PocketSphinx only starts recording speech after the keyword is spoken or received. This reduces false positives and improves the accuracy of the system. CMUSphinx has several acoustic and language models such as U.S. English, German, Spanish, Indian English, etc. This allows PocketSphinx to adapt to the accent of the user. In this case, U.S. English was used in conjunction with a list from a dictionary file. The file had words representing electrical components that are to be switched on and off, and explains how the user should pronounce the words.

Fig. 6.3 shows the Raspberry Pi connected to a microphone using a USB webcam. Table 6.1 depicts the phonetic used by PocketSphinx for this research. The table contains sequences of phonemes and their respective vocabulary words that we are interested in recognizing. The dictionary file can be generated on the CMUSphinx website and be used when a narrow search of pertinent words is preferred instead of using an entire language model [14]. The dictionary file guides both PocketSphinx and the user on how words should be

TABLE 6.1 Vocabulary Word and Respective Sequences of Phonemes

Vocabulary Word	Sequences of Phonemes
CLOSE	K L OW S
CLOSE(2)	K L OW Z
FAN	F AE N
GARAGE	G ER AA ZH
LIGHT	L AY T
OFF	AO F
ON	AA N
ON(2)	AO N
OPEN	OW P AH N
TIME	T AY M
TV	T IY V IY
TV(2)	T EH L AH V IH ZH AH N

pronounced. That makes it easy for PocketSphinx to recognize spoken words, and reduce the words that it has to search. This improves both performance and rate of which spoken commands are converted into text. The words below represent operations the system is able to do. For example, if the user wants to switch on a light, the user would say "LIGHT" which is "L AY T" according to the PocketSphinx dictionary.

6.3.2 System Automation

The overall objective of this home automation system is to integrate with existing infrastructure in a home and have little implications on the usage of existing home appliances. Because of this, the system employs infrared as a form of communication with appliances in a home. This approach enables the system to control most types of appliances without requiring the appliances to be smart or network enabled. However, the system can also be integrated with smart devices by mapping a voice command with the appropriate action.

To control appliances the system uses Linux Infrared Remote Control (LIRC). LIRC enables the system to send and receive infrared signals. The system uses LIRC to synchronize home appliance remote controls to the automation system, and to transmit infrared signals to the appliances as if it the signals were coming from the remote controls. LIRC enables the system to have command over multiple devices that can be controlled using infrared remotes. In this case, LIRC was synchronized with an air conditioner remote to turn on/off whenever the word "FAN" is said. This was done by pointing the remote control and pressing the power button to get the infrared signal of the button. Python can be used as the scripting language to match the text with an infrared signal to switch on the air conditioning unit every time the word "FAN" is detected in a spoken command.

6.3.3 Other Applications

IoT devices have evolved over the years, which means there are so many applications of IoT in our everyday lives that we don't realize we are utilizing. IoT can enhance an automation system by adding various sensing components to the system. This part of the chapter looks at enhancements that can be added to the system to automate other tasks in a home. The first is the addition of a light and ultrasonic sensor to the system. An ultrasonic sensor transmits ultrasonic sound waves and can be used to detect the presence of motion in a room. This sensor can be used to automate lighting and other appliances in a home. For example, if the system detects that there is no motion in a room, it will determine what appliance are on that do not need to be because no one is in the room. After this determination, the system can then decide to turn all of them off. This can include a TV or lights in a room. Of course, it can be overridden with user voice commands. If the user walks in a room the system can turn the appliances back on, unless otherwise specified by the user.

With a light intensity sensor, working in conjunction with the system, it will be able to keep lights on in a room only when necessary. If the light intensity in the room is high, the system will determine that there is no need to have the lights on and turn them off. Similarly, if there is no motion in a room for a specified amount of time, the system can switch off the light until such a time that there is motion detected.

A soil moisture sensor can also be added to the system to assist in automated irrigation for small plants in a home [15]. The sensor can connect to the system and operate autonomously to give the user control over plant irrigation. Automation will also require monitoring different aspects of the system to ensure that it is operating as expected. For this reason, we can develop a local website and mobile application that can show the status of different components of the automation system. The mobile application can also be used to control the devices in the automation system.

6.3.4 Results and Discussion

Fig. 6.4 shows a sample of the system processing speech. In this output we, the system attempts to recognize the word "FAN" from which is contained in the dictionary file. Here is the output of one speech recognition.

The results show a speaker saying the word "FAN". The results show that 2063 words were recognized at a rate of 6 words per frame, 41,553 senones evaluated at 121 senones per frame and taking a combined total time of 0.186 extract performance data (xRT). This indicates that the system is able to convert speech to text when it is spoken by users with different accents based on their ethnic backgrounds. With the use of LIRC, the spoken command can be

```
INFO: cmn_live.c(120): Update from < 25.69  2.53  1.88  1.01  3.06  1.39  1.16  1.51  3.55  0.49
 0.48 -0.45  1.88 >
INFO: cmn_live.c(138): Update to   < 27.05  4.52  0.45 -0.80  2.87  2.79  2.70  1.37  4.72  0.48
-0.48 -1.96  0.43 >
INFO: ngram_search_fwdtree.c(1550):     2229 words recognized (5/fr)
INFO: ngram_search_fwdtree.c(1552):    42794 senones evaluated (104/fr)
INFO: ngram_search_fwdtree.c(1556):    31337 channels searched (75/fr), 2453 1st, 27176 last
INFO: ngram_search_fwdtree.c(1559):     2686 words for which last channels evaluated (6/fr)
INFO: ngram_search_fwdtree.c(1561):     1339 candidate words for entering last phone (3/fr)
INFO: ngram_search_fwdtree.c(1564): fwdtree 1.76 CPU 0.425 xRT
INFO: ngram_search_fwdtree.c(1567): fwdtree 5.77 wall 1.397 xRT
INFO: ngram_search_fwdflat.c(302): Utterance vocabulary contains 7 words
INFO: ngram_search_fwdflat.c(948):     2476 words recognized (6/fr)
INFO: ngram_search_fwdflat.c(950):    48936 senones evaluated (118/fr)
INFO: ngram_search_fwdflat.c(952):    44437 channels searched (107/fr)
INFO: ngram_search_fwdflat.c(954):     3678 words searched (8/fr)
INFO: ngram_search_fwdflat.c(957):     1246 word transitions (3/fr)
INFO: ngram_search_fwdflat.c(960): fwdflat 0.76 CPU 0.185 xRT
INFO: ngram_search_fwdflat.c(963): fwdflat 0.77 wall 0.188 xRT
INFO: ngram_search.c(1250): lattice start node <s>.0 end node </s>.395
INFO: ngram_search.c(1276): Eliminated 1 nodes before end node
INFO: ngram_search.c(1381): Lattice has 494 nodes, 1184 links
INFO: ps_lattice.c(1380): Bestpath score: -8620
INFO: ps_lattice.c(1384): Normalizer P(O) = alpha(</s>:395:411) = -427516
INFO: ps_lattice.c(1441): Joint P(O,S) = -489874 P(S|O) = -62358
INFO: ngram_search.c(872): bestpath 0.01 CPU 0.002 xRT
INFO: ngram_search.c(875): bestpath 0.01 wall 0.002 xRT
FAN FAN
INFO: continuous.c(275): Ready....
```

FIG. 6.4 Sample of the system processing speech.

matched to a corresponding infrared signal that the system can transmit with an infrared transmitter to turn the air condition unit on/off. With additional hardware like an ultrasonic sensor and light intensity sensors the system can also control lighting in rooms or manage appliances by turning them on/off based on the presence or absence of a person or lighting in a room.

6.4 Significance/Contribution

Home appliances are also using wireless technologies and can be accessed by radio communication and also accessed from outside the home by the Internet, which will make life easier. Although this technology is advancing in many other sectors like road traffic, hotels, etc., implementing it in homes, especially in developing countries, has not been very successful. That's because of the high cost for the hardware to enable a home to be fully automated and monitored by using human speech. Most people in homes rely on caregivers to assist them in doing everyday tasks, but given the choice, some people might appreciate the convenience of automation in a home. The approach shown in this chapter aims at enabling people to be able to automate processes in the home by using voice commands. This will reduce the reliance on caregivers by elderly and physically challenged people. The proposed system approach shows the possibility of using an STT engine to convert spoken commands into text, and then using the text as a form of commands that the system will use to transmit infrared signal to the appliance.

The system uses PocketSphinx to capture speech signal, and then convert the speech to text. The system then analyses the converted text to determine if it contains any known commands. The system also employs LIRC in conjunction with the infrared receiver and transmitter. A component of the system is used to train the system to use the correct infrared to transmit to an appliance. This training is done using an infrared receiver to synchronize the system with certain buttons of the remote control of the home appliance. Once the system is synchronized, the system uses the infrared transmitter to control the appliance as if the signal was being transmitted from the appliance remote control.

This innovative approach to home automation enables the system to integrate into homes by being able to control home appliances without requiring the additional purchase of other hardware or new home appliances, which makes it cost-effective.

Limitation: The proposed Smart Home Automation System cannot be used by speech-impaired people.

6.5 Conclusion

This chapter presented a VAD home automation system that uses low-cost hardware to enable users to automate processes in a home. This automation can be beneficial to users that rely on caregivers to do day-to-day tasks like controlling appliances in homes. These users include the elderly, physically challenged people and other users who have the need to automate processes and tasks in a home. The system uses a STT engine to convert spoken commands to text, and uses the text to transmit a corresponding infrared signal to an appliance as if the infrared signal was coming from a remote control of the appliance. The system can further be enhanced by incorporating other sensors and actuators like ultrasonic sensors, light intensity sensors, which can make the system act autonomously in controlling other appliances in a home. This indicates that we can all benefit from home automation approaches that can integrate with our current appliances.

References

[1] Namibia Statistics Agency, Namibia 2011 Census Disability Report, Retrieved from:https://cms.my.na/assets/documents/Namibia_2011_Disability_Report.pdf, 2016.

[2] M. Asadullah, A. Raza, An overview of home automation systems, in: Proceedings of 2nd International Conference on Robotics and Artificial Intelligence, IEEE Press, Rawalpindi, Pakistan, 2016.

[3] M. Price, J. Glass, A.P. Chandrakasan, A low-power speech recognizer and voice activity detector using deep neural Networks, IEEE J. Solid State Circuits 53 (2018) 66–75.

[4] J. Kim, B. You, Fault detection in a microphone array by intercorrelation, Power 58 (6) (2011) 2568–2571.

[5] Fuzzy Logic and Artificial Neural Networks, Development of Robust VAD Schemes for Voice Operated Switch Application in Aircrafts, (2016) pp. 191–195.

[6] E. Chuangsuwanich, S. Cyphers, J. Glass, S. Teller, Spoken command of large mobile robots in outdoor environments, Artificial Intell. (2010) 294–299. https://ieeexplore-ieee-org.eresources.nust.na/stamp/stamp.jsp?tp=&arnumber=5700869&tag=1.

[7] S.S. Thanu, A.R. Uthra, A smart home energy management system in intelligent building, Int. J. Eng. Dev. Res. 3 (2) (2015) 400–403.

[8] S. Folea, D. Bordencea, C. Hotea, H. Valean, Smart home automation system using wi-fi low power devices, in: Proceedings of 2012 IEEE International Conference on Automation, Quality and Testing, Robotics, IEEE Press, Cluj-Napoca, Romania, 2012.

[9] S. Soumya, M. Chavali, S. Gupta, Internet of things based home automation system, in: Proceedings of IEEE International Conference on Recent Trends in Electronics, Information & Communication Technology, IEEE Press, Bangalore, India, 2017.

[10] N. Xu, C. Wang, J. Bao, Voice activity detection using entropy-based method, in: Proceedings of 9th International Conference on Signal Processing and Communication Systems, IEEE Press, Cairns, Australia, 2016.

[11] N. Dey, A.S. Ashour, Challenges and future perspectives in speech-sources direction of arrival estimation and localization, in: Direction of Arrival Estimation and Localization of Multi-Speech Sources, Springer, Cham, 2018, pp. 49–52.

[12] CMUSphinx, Versions of Decoders, PocketSphinx, 2018. https://cmusphinx.github.io/wiki/versions/(accessed 26.09.18).

[13] A.M. Johansson, E.A. Lehmann, S. Nordholm, Real-time implementation of a particle filter with integrated voice activity detector for acoustic speaker tracking. in: IEEE Asia-Pacific Conference on Circuits and Systems, Proceedings, APCCAS, 2006, pp. 1004–1007, https://doi.org/10.1109/APCCAS.2006.342257.

[14] Sphinx Knowledge Base Tools, Sphinx Knowledge Base Tool—VERSION 3, 2018. http://www.speech.cs.cmu.edu/tools/lmtool-new.html (accessed 26.09.18).
[15] Jiankai.li, Grove—Moisture Sensor User Manual, https://www.mouser.com/ds/2/744/Seeed_101020008-1217463.pdf, 2015, (accessed 26.09.18).

Further Reading

J. Ling, S. Sun, J. Zhu, Speaker recognition with VAD, in: Proceedings of Second Pacific-Asia Conference on Web Mining and Web-Based Application, Computer Society Press, California, USA, 2009.

J. Mayer, IoT architecture for home automation by speech control aimed to assist people with mobility restrictions, in: Proceedings of ICOMP'17—The 18th Int'l Conf. on Internet Computing and Internet of Things, CSREA Press, Nevada, USA, 2017.

Chapter 7

Speech Summarization for Tamil Language

A. NithyaKalyani* and S. Jothilakshmi†
*Department of Computer Science and Engineering, Annamalai University, Chidambaram, India,
†Department of Information Technology, Annamalai University, Chidambaram, India

7.1 Introduction

Speech recognition is the methodology, in which a computer recognizes the speech that is given as input to the computer program, and provides the textual output of the speech. It has a few more applications like speaker identification, structure identification, speech analysis (recognizing the emotion, nature of speech), etc. Speech summarization is the process of retrieving the essential information from speech files and producing the extracted data in a concise form to benefit the end users. Speech summarization uses the speech recognition technique and applies natural language processing algorithms to summarize the textual results obtained from the recognition system. Approaches for summarization can be either extractive or abstractive. Extractive summarization aims at identifying the salient information that is then extracted and grouped together to form a concise summary. Abstractive summary generation rewrites the entire document by building internal semantic representation, and then a summary is created using natural language processing. Other dimensions of summarization [1] include:

- Indicative versus informative—An indicative summary contains only the description of the spoken document and not the informative content. For example, the title page of books or reports. An informative summary contains the informative part of the original document. For example, research articles where the essential part of research is discussed.
- Generic versus query-driven—In the query-driven approach, based on the given query, the information that is closely connected to the query is extracted. In the generic approach, the overall concept discussed in the document is presented.
- Single versus multidocument—The summary can be generated from a single source document (or) from the multiple sources of a document.

Intelligent Speech Signal Processing. https://doi.org/10.1016/B978-0-12-818130-0.00007-6

- Single versus multiple speakers—The summary is generated from the information presented by a single speaker or from multiple speakers where the speaker's details are also incorporated in the summary.
- Text-only versus multimodal—The summarization result can be presented either as text or as a speech.

Based on number of speakers and speaking style, there are various forms of speech from which a summary can be generated. They include:

- Broadcast news
- Lecture
- Public speaking
- Interview
- Telephone conversation
- Meeting

7.2 Extractive Summarization

Summarization using extractive approaches have demonstrated a growing popularity in the past decades. Both unsupervised and supervised approaches have been explored for speech summarization.

7.2.1 Supervised Summarization Methods

The summarization problem, in general, is handled as a two-class, sentence-classification problem by the supervised machine learning algorithms. Here, the sentences are classified as summary class and nonsummary class [2]. To characterize a spoken sentence, say S_i, a set of indicators such as structural features, relevance features, acoustic features, lexical features, and discourse features were used. The classifier takes the corresponding feature vector say X_i of S_i as input, and based on the output classification score, the sentence will be selected as either part of the summary or not. Thus by constructing a ranking model that is used to allot a classification score to each sentence included in a summary, the most relevant and salient sentences are ranked and preferred based on the scores. The summarizer iteratively concatenates the sentences to the summary until the desired summarization ratio is achieved. The higher the score, the more the performance of the summarizer will be improved. While training the summarizer, it gives many errors of classifying sentences incorrectly; this is bridged using heuristic methods like sampling and resampling [3].

Supervised methods require a huge amount of spoken dataset and their equivalent manual summaries for training purposes and to prepare the classifiers. Creating the summaries manually for all available spoken data requires more number of human resources, and also it consumes more time. The main problem with supervised summarizers is that they restrict their abstraction skill, and so it may not be easily applicable for a new task or domain. In addition, the supervised summarizers suspect that the content or the information provided by each sentence in the spoken document does not depend on each other. Hence, the sentences are treated as individual sentences and are classified, accordingly. Supervised summarizers do not rely on the dependence relationship among every sentence [4]. There are different machine learning algorithms such as Gaussian mixture model [5], Bayesian classifiers [6], support vector machines (SVM) [7], conditional random fields [8], ranking SVM [9], global conditional log-linear model [9], deep neural network [9], and perceptron [10].

7.2.2 Unsupervised Summarization Methods

Unsupervised summarization methods depend on some statistical evidence such as number of occurrences of words in each and every sentence as well as in the entire document. Unsupervised methods do not depend on manual annotations of training data. They conceptualize the sentences based on their weight of importance and weight of relationship with the document. This is done by classifying the sentences as the nodes and the link between them as their lexical relationships. Since the scores are based on probability, the results of unsupervised summarizers are worse than supervised. However, they are domain-independent and easy to implement.

In the vector space model (VSM), the complete document, including every sentence, is represented using a vector format [11]. Here, each dimension represents a collection of quantitative data. For example, the term frequency score (*tf*) and inverse document frequency score (*idf*) are multiplied and the resultant score will be associated with a word or a sentence in the document. The statements with the prominent relevance scores to the entire spoken document are considered to be incorporated in the summary generation process [12].

Latent semantic analysis (LSA) [13] makes use of vectors to represent every statement in a spoken document. Singular value decomposition (SVD) is implemented on a set of matrices representing the words and sentences of a spoken document, and so vectors are created. The more dominant latent semantic concepts are represented by wide-ranging singular values in the right singular vectors. Thus the values in right singular vectors are treated as salient sentences and those sentences are given more preference for summary generation.

Dimension reduction (DIM) is another LSA-based technique [14] where the relevance score of each sentence is calculated, and it depends on the normalized vector in a lower m-dimensional latent semantic space. Then the sentences with higher scores are identified and considered for inclusion in the summary generation.

In case of maximum margin relevance (MMR) [15], sentences are selected iteratively based on two rules: (1) whether the sentence is better related to the context of the complete speech transcript compared with other sentences, and (2) whether the sentence is less similar than other sentences to the already selected set of sentences [16]. Thus in addition to selecting suitable sentences for the summary, it also considers more concepts to be covered in the summary.

Graph-based techniques, including LexRank [17], TextRank [18], Markov random walk [19], consider spoken content to be condensed as a network of sentences. Here, the node and the edges between each node, represent a sentence and a relationship between each sentence. The weight or values associated with the edges represent the lexical similarity relationship between each sentence. Document summarization in general, not only considers the local features of each sentence, but also considers the global structural information present in a conceptualized network. Daumé et. al. [20] explored the use of probabilistic models to collect the similarity details among sentences in the speech transcript.

Chen et al. [21] conducted a summarization task in a completely unsupervised manner by utilizing the framework of probabilistic ranking for summarizing speech files. In this framework, the length of space between a statement and a document model is determined and most salient statements are selected based on either the computed scores or the likelihood of a model generating the summary of a spoken content. Unsupervised summarization approach is independent of domain and is effective in terms of effortless execution compared to supervised summarization approach.

7.3 Abstractive Summarization

Abstractive summary generation rewrites the entire document by building internal semantic representation, and then uses natural language generation to create a summary. Table 7.1 depicts the broad classification of abstractive summarization techniques [22].

7.3.1 Structured Approach

The structured approach converts the essential facts available in the spoken report into a particular form through subjective strategy [23]. The categories of structure-based approaches are described below.

TABLE 7.1 Abstractive Summarization Techniques

Structured approach	Tree based technique
	Template based technique
	Ontology based technique
	Lead and body phrase technique
	Rule based method
Approach based on semantics	Multimodal semantic technique
	Information item based technique
	Semantic graph based technique

7.3.1.1 Tree-Based Technique

In this approach, the information in the document is represented using a dependency tree. Several algorithms such as theme intersection algorithm, or one that makes use of local alignment across pair of parsed sentences are used to select the important information from the document for summary generation. Related literature using this method shall be referred in [24, 25].

7.3.1.2 Template-Based Technique

Here, the document is represented using a template. The framework of linguistic patterns is used to locate a piece of information from the text document and they are mapped to a template slot. The identified piece of information acts as an indicator for the content selection to generate summary. Related literature using this method shall be referred in [26].

7.3.1.3 Ontology-Based Technique

Research on summarization has been improved by making use of the ontology concept. Data available on the Internet are grouped based on their similarity, so it is domain related. The way of organizing the knowledge is domain dependent and ontology helps in describing it in a better way. Detailed study on this technique shall be referred in [27].

7.3.1.4 Lead and Body Phrase Technique

In this technique, operations on phrases such as insertion and substitution form the base where the head and body of the sentences are analyzed for same syntactic head piece. The sentences are searched for triggers, and using the similarity metric, the sentences with maximum phrase are recognized. If rich information is available in the body phrase, and if it has its equivalent phrase, then substitution is performed. If the body phrase has no correspondent, then insertion is performed. Substitution and insertion into the body phrase has information rich context in the summary. Related literature using this method shall be referred in [28].

7.3.1.5 Rule-Based Technique

In this technique, a rule-based information extraction module along with selection heuristics is used to select a sentence for summary generation. The extraction rules are formed using verbs and nouns that have similar meaning. The summary generation module makes use of several candidate rules to provide the best summary. Related literature using this method shall be referred in [29].

7.3.2 Semantic-Based Approach

In this approach, the linguistic data is utilized to identify the phrases of nouns and verbs [30]. The techniques under this approach include.

7.3.2.1 Multimodal Semantic Technique

In this method, a semantic model is used to identify the relationships among the content of the document. The content of the document can be either text or images. The important information identified in the document is rated and based on some measures, the informative sentences are expressed to form a summary. The summary generated using this technique has better quality, because all the graphical and salient information is covered in the summary. Related literature using this method shall be referred in [31].

7.3.2.2 Information Item-Based Technique

In this method, the information required for summary generation is selected from the overall representation of source file rather than from the original document. Information item is the abstract way of representation, and it is the smallest unit of logical information present in the text document. Information item is extracted through syntactic analysis and the sentences are ranked using the average document frequency score. Then a summary is generated, which includes all the characteristics of date and location. Finally, a logical and information rich summary is generated. Related literature using this method shall be referred in [32].

7.3.2.3 Semantic Graph-Based Technique

This technique makes use of a rich semantic graph, which is used to represent the verbs and nouns in the document as graph nodes, and the edges between the nodes represent the semantic relationship, and the topological relationship between the verbs and nouns. Later, some heuristic rules are applied to reduce the rich semantic graph so as to generate an abstractive summary. The advantage of this technique is that the summary sentences are grammatically corrected, scalable, and less redundant. Related literature using this method shall be referred in [33].

7.4 Need for Speech Summarization

The natural way of information exchange among human beings is speech, where the message along with the prosodic information, such as emotion, is also conveyed. When the speech (e.g., lectures, presentations, news) is recorded as an audio signal, the complete information is conveyed to the listener, but it is difficult for the listener to quickly recall the information delivered. Therefore transcribing speech became the mandatory part, and the speech recognition system played a vital role to generate a transcript for the given speech documents.

Though the speech recognition system provided accurate results for the input speech that is read from a text (broadcast news read by an anchor), the efficiency of transcribing continuous speech is still minimal.

Spontaneous speech varies from the written text and it includes redundant information due to the presence of breaks and irregularities in the speech. The transcripts of spontaneous speech may include redundant and irrelevant information due to the fillers, word fragments, and recognition errors. Therefore the transcripts for spontaneous speech alone will not provide the efficient information that is required by the end user. Instead, the most important and relevant information shall be extracted from the spontaneous speech transcripts. and it can be rendered in the form of summary of the speech files. The summary of recorded speech saves time to review and recall the information present in the speech and thus improves the efficiency of document retrieval. The summary, in turn, can be in the form of either speech or text.

7.5 Issues in the Summarization of a Spoken Document

- Identify utterances: An utterance is a way of conveying information and it differs by speaker and language, but it has no effect on the content.
- Human variations: Different people tend to choose different sentences and this will not lead to quality in the generation of a summary.
- Semantic equivalence: One or more sentences may convey the same information, thus it is not advisable to use only the sentences as the selection unit for summary generation. For example, news and multidocument summarization.
- Another issue for automatic speech summarization is how to deal with recognition results, including word errors. Handling word errors is a fundamental aspect for successfully summarizing transcribed speech. In addition, since most approaches extract information based on each word, approaches based on a longer phrase, or compressed sentences are required for extracting messages in speech.
- The major issue in speech recognition is that the transcripts generated by the recognition module may not be linguistically correct, and it is due to the recognition errors. Hence, it is necessary to develop a technique to automatically summarize the speech to cope with such problems.

7.6 Tamil Language

Tamil language is a classical Indian language and is extensively spoken in Tamilnadu, which is the southern state of India. Tamil language is based on syllables. It includes 18 consonants (மெய்யெழுத்து; meyyeḻuttu), 12 vowels (உயிரெழுத்து; uyireḻuttu) and one another unique letters "aytham (அக்கு ∴)". The vowels are categorized as five short vowels and seven long vowels that include two diphthongs. The 18 consonants are grouped as hard, soft, and medium with 6 consonants in each group. The categorization of consonants

TABLE 7.2 Syllabic Rules

Pattern	Example
Short vowel + long vowel + consonant(s)	கனா (Kaṉā)
Short vowel + long vowel	விழா (Viḻā)
Short vowel + short vowel + consonant(s)	களம் (Kaḷam)
Short vowel + short vowel	கல (Kala)
Short vowel + consonant(s)	பல் (Pal)
Long vowel + consonant(s)	கால் (Kāl)
Long vowel	வா (Vā)
Short vowel	க (Ka)

depends on the point of articulation, and the vowels are produced by varying the position of the tongue in different parts of the mouth. The combination of both consonants and vowels produce a syllable. The consonants in general are represented with a dot on top of the symbol (க் /k/), and it is mingled with a vowel (அ /a/) to form a syllable (க /ka/) [34, 35].

One of the unique features in the Tamil language is a prosodic syllable (*asai*) [36], and the pronunciations are based on the prosodic syllable. The representation of the prosodic syllable in Tamil language is categorized as *Ner Asai* and *Nirai Asai*. The rules for the prosodic syllable constitute eight patterns and are described in Table 7.2.

7.6.1 Tamil Unicode

A universal character encoding scheme is said to be a Unicode and it is available for written characters and text. For English language, ASCII characters are used and in the same way, for Tamil language. Unicode characters are used. The effectiveness of using Unicode is that it is platform independent and program independent. It can be used to encode multilingual text, and acts as a basis for global software. The Tamil Unicode range is U+0B80 to U+0BFF and its decimal value range from 2944 to 3071. The Unicode characters are comprised of 2 bytes in nature. For example, consider the word வணக்கம். Its corresponding Unicode and decimal value is presented in the Table 7.3.

7.7 System Design for Summarization of Speech Data in Tamil Language

Exhaustive research on text summarization has already been done for English and other foreign languages, whereas it is still lacking for Indian languages.

TABLE 7.3 Unicode for the Word வணக்கம்

Tamil Letter	Unicode	Decimal Value
வ (VA)	U+0BB5 TAMIL LETTER VA	2997
ண (NNA)	U+0BA3 TAMIL LETTER NNA	2979
க (KA)	U+0B95 TAMIL LETTER KA	2965
· (puḷḷi—tamil sign virama)	U+0BCD TAMIL SIGN VIRAMA	3021
க (KA)	U+0B95 TAMIL LETTER KA	2965
ம (MA)	U+0BAE TAMIL LETTER MA	2990
· (puḷḷi—tamil sign virama)	U+0BCD TAMIL SIGN VIRAMA	3021

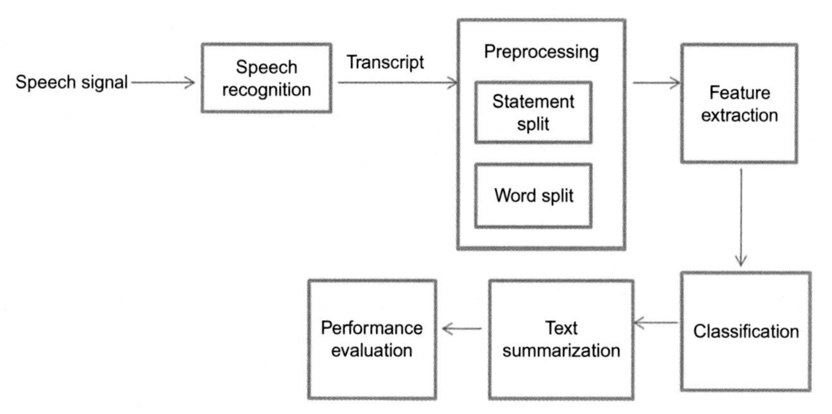

FIG. 7.1 Architecture of spoken document summarization for Tamil language.

Text summarization for Indian language has evolved and research work is in progress. Speech summarization for Tamil language is a domain that is not explored so far, and it has to be studied to help people who are hearing impaired. Fig. 7.1 describes the overview of steps involved in summarization of a spoken document for Tamil language.

The major steps involved are: speech recognition, feature extraction from the speech transcript, classification, summarization, and performance evaluation.

7.7.1 Speech Recognition Techniques

Speech recognition is the way to translate the input speech signal into its corresponding transcript [37]. Generations of transcripts from the input speech signal is a challenging task when it comes to native languages like Tamil, because

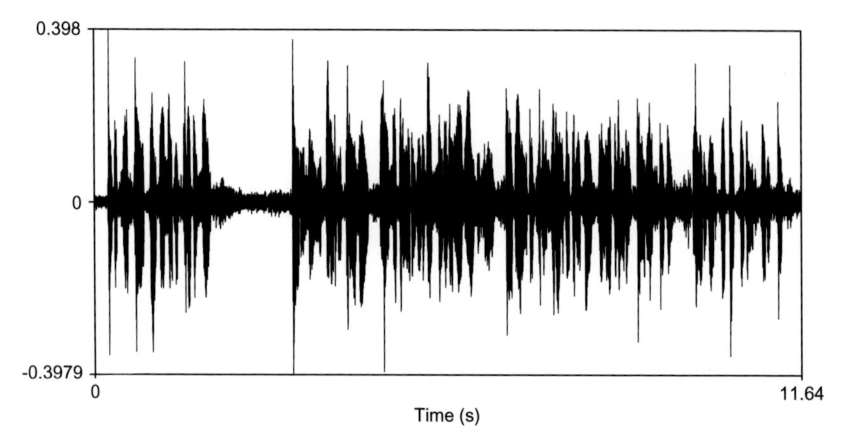

FIG. 7.2 Speech signal for LDCIL_Speech_Annotation_Sample_Tamil\LDCIL_Speech_Annotation_Sample_Tamil\T1-0001-009\T1-0001-009_mono.wav.

of the variations in accents and dialects. Research is ongoing in Tamil speech recognition to solve issues and challenges to develop an effective recognition system Fig. 7.2 shows the speech signal of a sample WAV file obtained from the Linguistic Data Consortium for Tamil language. Fig. 7.3 shows its corresponding transcript.

The classical speech recognition system consists of two phases [38]; preprocessing and postprocessing. The preprocessing step focuses on the feature extraction part, where the features are extracted from the input speech signal. The purpose of feature extraction from input speech waveform is to characterize it at a lower information rate for further exploration. The widely used feature extraction techniques are [39]: mel-frequency cepstral coefficients (MFCC), linear predicting coding (LPC), linear predictive cepstral coefficients (LPCC), perceptual linear predictive coefficients (PLP), wavelet features, auditory features, and relative spectra filtering of log domain coefficients (RASTA).

The components in the postprocessing step are: acoustic models, a pronunciation dictionary and language model. The statistical way of representing the

"அவற்றை கேட்ட நமது ப்ரதமர் அரசியல் தீர்வுக்காண நமது அரசு என்ன செய்ய இயலுமோ அதை அவசியம் செய்வதாக உறுதி அளித்தார் என்பது வரவேற்கதக்கதொரு திருப்பம்"

"Avaṟṟai Kēṭṭa Namatu Pratamar Araciyal Tīrvukkāṇa Namatu Aracu Eṉṉa Ceyya Iyalumō Atai Avaciyam Ceyvatāka Uṟuti Aḷittār Eṉpatu Varavēṟkatakkatoru Tiruppam"

FIG. 7.3 Speech transcript of LDCIL_Speech_Annotation_Sample_Tamil\LDCIL_Speech_Annotation_Sample_Tamil\T1-0001-009\T1-0001-009_mono.wav.

feature vector generated from a speech signal is referred to as acoustic modeling. Acoustic modeling is used to explore the representation of words and sentences, which is considered a relatively larger speech unit in a spoken document. The pronunciation dictionary is a resource that is language dependent, and it includes all possible words that the speech recognition module can understand. Also, the dictionary contains the details about the different ways of pronouncing the words. The language model is used to explore the possibilities of word sequence and how frequently they occur together. The language model is also used to regulate the searching process for the identification of a word. Table 7.4 shows the advantages and disadvantages of each speech recognition technique [40].

7.7.2 Isolated Tamil Speech Recognition

As an initial stage in Tamil speech recognition, a sample set of 15 Tamil words are uttered by 4 different people (1 male and 3 female). Each word is repeated 10 times to have 10 different utterances. Therefore the total size of the dataset is 600 ($15 \times 4 \times 10$). Audacity software is used to record the utterances and MATLAB software is used to perform the experiment. Here, 60% of the spoken data is used for training purpose and the remaining 40% of the spoken data is used for testing purposes. The state-of-the-art speech recognition techniques were used to recognize the same speech samples and their performance is compared based on the word error rate, word recognition rate and real-time factor [40]. Fig. 7.4 shows the word level accuracy and average time taken during the training and testing process.

From Fig. 7.4, it is proven that HMM and DTW techniques provide better results in comparison to other state-of-the-art methods. Also, it is shown that the statistical approaches works better for isolated word recognition, and in the case of speaker independent applications, machine learning approaches perform better.

7.7.2.1 Related Work on Tamil Speech Recognition

Lakshmi et al. [41] proposed the syllable-based continuous speech recognition system. Here, the group delay-based segmentation algorithm is used to segment the speech signal in both training and the testing process and the syllable boundaries are identified. In the training process, a rule-based text segmentation method is used to divide the transcripts into syllables. The syllabified text and signal are used further to annotate the spoken data. In the testing phase, the syllable boundary information is collected and mapped with the trained features. The error rate is reduced by 20% while using the group delay based syllable segmentation approach, and so the recognition accuracy is improvised.

Radha et al. [42] proposes the automatic the Tamil speech recognition system by utilizing the multilayer feed-forward neural network to increase the recognition rate. The author-introduced system, eliminates the noise present in the

TABLE 7.4 Comparison of Speech Recognition Techniques

Techniques	Advantage	Disadvantage
Dynamic time warping	Execution is made easy, and also random time warping can be modeled.	A rule-based approach is used to measure distances and the warping paths. The best solution or the convergence is not guaranteed. When the size of the vocabulary grows, it doesn't scale well and shows poor performance when the environment changes.
Hidden Markov model	HMM makes use of positive data and so, it is easily extendable. Though the vocabulary size grows, the time taken for the recognition process is minimal.	It requires large number of parameters and the data required for training is also large. The likelihood of examining data instances from other classes is reduced.
Gaussian mixture model	The probability estimation is perfect and the classification yields the best solution.	Time taken for complete recognition process is not reduced comparatively.
Multilayer perceptron	Based on the discriminative criteria, learning is performed. Continuous functions shall be approximated using a basic structure. The complete idea about the type of input is not required, and it produces some reasonable recognition results for the test case input that has not been taught before.	For a continuous speech recognition task, the performance is not good. Though the recurrent structures are defined, MLP lacks in building a speech model.
Support vector machine	Robustness is increased and the training process is easy. In the case of high dimensional data, scaling is relatively high and it does not require local optimality.	SVM requires a good kernel function.
Decision trees	It is easy to understand and manipulate. Also, it will work along with other decision techniques.	The complexity increases when the data size is increased.

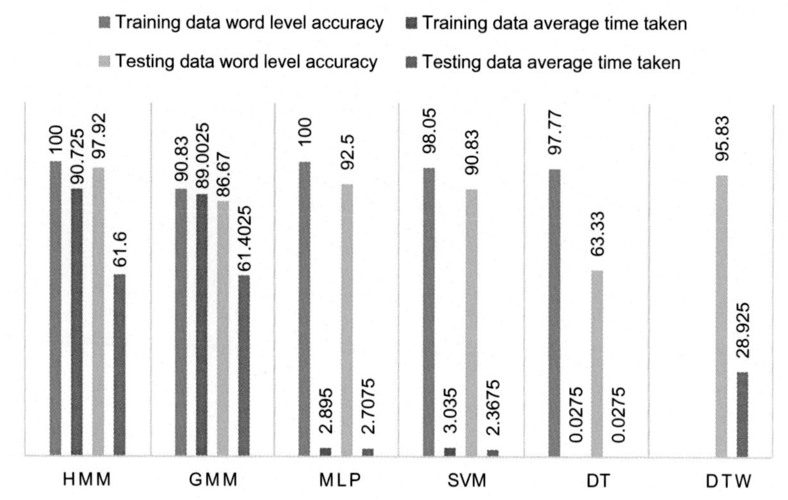

FIG. 7.4 Performance analysis of speech recognition techniques.

input audio signal by applying the preemphasis, median, average, and Butterworth filter. Then the linear predictive cepstral coefficients features are identified and extracted from the preprocessed signal. The extracted features are classified by applying the multilayer feed-forward network, which classifies the Tamil language efficiently. The performance of the system is analyzed with the help of the experimental results, which reduces the error rate and increases the recognition accuracy.

Alex Graves et al. [43] recognize the speech features by applying the deep neural network because it works well for sequential data. The network works are based on the long- and short-term memory process to analyze the interconnection between the speech features. The extracted features are classified in terms of the connectionist temporal classification process. The implemented system reduces the error rate up to 17.7%, which is analyzed using the TIMIT phoneme recognition database.

Gales and Young [44] implement the large vocabulary continuous speech recognition system for improving the recognition rate. The authors reduce the assumptions about the particular speech features, which are classified by applying the hidden Markov model. This model uses the various process like feature projection, discriminative parameter estimation, covariance modeling, adaption, normalization, multipass, and noise compensation process while detecting the speech feature. The proposed system reduces the error rate, and so the recognition rate is increased in an effective manner. Table 7.5 describes the performance of various speech recognition techniques and it demonstrates that the modified global delay function with the Gammatone wavelet coefficient approach yields the better recognition, comparatively.

TABLE 7.5 Comparison of Speech Recognition Techniques

Recognition Technique	Recognition Accuracy in Percentages
MFCC with HMM [45]	85
Mel-frequency cepstral coefficients (MFCC) with deep neural network (DNN) [46]	82.2
Gammatone cepstral coefficients (GTCC) with hidden Markov model (HMM) [47]	85.6
Gammatone cepstral coefficients (GTCC) with Deep Neural Network (DNN) [48]	88.32
Modified global delay function (MGDF) with Gammatone wavelet coefficient approach [49]	98.3
Syllable-based continuous approach [41]	80

7.7.3 Features Used for Summarization

Here, the feature classes such as lexical, structural, acoustic/prosodic, discourse, and relevance features are discussed [50]. They are used to predict sentences that will be extracted to generate a summary for the spoken document. Table 7.6 shows the comparison results of different features that are used to characterize a spoken document.

7.7.3.1 Acoustic/Prosodic Feature

Features that are extracted from raw speech signal are denoted as acoustic or prosodic features. They describes more about how things are said, than what is said. Table 7.7 shows the various forms of acoustic features. F0 features include first formant, second formant, and third formant features. For each feature, the maximum, minimum, difference, and average value of a spoken sentence are extracted. A change in pitch may be a topic shift. The RMS energy feature (maximum, minimum, difference, and average) is used to illustrate the higher amplitude that probably means a stress on the phrases. Duration represents the length of the sentence in seconds (end time–start time); a short or long sentence might not be important for summary. The speaker rate indicates how fast the speaker is speaking; the slower rate may mean more emphasis in a particular sentence.

7.7.3.2 Lexical Feature

It is used to represent the linguistic characteristics. Some of the lexical features include: Named entities in a sentence such as person, people, organization, total count of named entities, number of stop words in a sentence, number of words in

TABLE 7.6 Comparison of Various Features Used to Characterize Spoken Document and Their Inherent Sentences

Features Used	Tested On	Observations
Lexical, structural [53]	Text and speech	Good performance for text compared to speech
Acoustic, prosodic, lexical, structural, discourse features [54, 55]	Broadcast news	Together, the acoustic, lexical, structural and discursive give the best performance. Acoustic and structural perform well when speech transcription is not available. An acoustic feature is useful in retrieving important sentences from the abstract of English broadcast news.
Lexical, acoustic, prosodic [56]	Lecture speech	Lexical contributes more than acoustic. Acoustic and prosodic permit representation of rhetorical information to improvise the performance of summarization.
Lexical, prosodic, structural, acoustic [51]	Broadcast news, lecture speeches	The manner of speaking by anchors and reporters is the same over time in the news domain, but it varies with lecture speakers. Structural and acoustic features perform well even without a lexical feature for broadcast news. A lexical feature performs well for lecture speeches.
Acoustic, lexical [57]	Meetings	Normalization of acoustic features improves summarization performance compared to lexical.
Acoustic feature [58]	Lecture speech	Speaker normalized acoustic feature improves performance.
Acoustic, lexical, structural [59]	Presentation speech	Acoustic and structural property produce a good performance and propose that the quality of speech summarization can be improvised without the extreme need of accurate results in speech recognition.

Continued

TABLE 7.6 Comparison of Various Features Used to Characterize Spoken Document and Their Inherent Sentences—cont'd

Features Used	Tested On	Observations
16 indicative features including lexical, prosodic, relevance and structural feature. [60]	Broadcast news	Relevance feature in isolation can achieve the best performance. Performance of prosodic feature is superior to lexical feature since it is less sensitive to the effect of imperfect speech recognition.
Acoustic, lexical, and structural features [61]	Broadcast news along with transcription obtained from LDC [62]	Performance of lexical feature is best compared to acoustic and structural feature. Also, the performance of ROUGE metric is better while using full feature set.

TABLE 7.7 List of Acoustic Features

Feature Name	Feature Description
Duration *I*, Duration *II*	Time and average phoneme duration of a sentence
Speaking rate	The average syllable duration
Energy value E*I*, E*II*, E*III*, E*IV*, E*V*	The minimum and maximum energy value, then its difference, mean, and slope of the energy
F0 formants F0*I*, F0*II*, F0*III*, F0*IV*, F0*V*	The minimum and maximum F0's value, then its difference, mean and slope of F0

previous and next sentence, bigram language model scores, and normalized bigram scores [51]. A lexical feature set contains eight features. These features are reported in Table 7.8. All lexical features are extracted from the manual transcriptions or ASR transcriptions.

$$\text{term frequency}\,(tf) = \frac{n_i}{\sum_k n^k} \tag{7.1}$$

where the numerator n_i refers to the number of occurrences of the examined words, and the denominator represents the number of occurrences of all words in a spoken document.

TABLE 7.8 List of Lexical Features

Feature Name	Feature Description
Len*I*	Total count of named entities
Len*II*	Count of words in previous sentence
Len*III*	Count of words in next sentence
Len*IV*	Count of stop words in a sentence
Len*V*	Score of a bigram language model
Len*VI*	Score of a normalized bigram
TFIDF	Term frequency inverse document frequency estimated using *tf* idf* as mentioned in Eqs. (7.1) and (7.2)
Cosine	Cosine similarity measure

$$\text{inverse document frequency}\,(idf) = \log \frac{|D|}{|(d_i \supset t_i)|} \tag{7.2}$$

Here, $|D|$ refers to the total count of sentences in the spoken document to be considered and $(d_i \supset t_i)$ refers to the number of sentences where the word t_i occurs.

7.7.3.3 Part-of-Speech Tagging

Part-of-speech (POS) tagging [52] is a technique to label every word in a sentence. It is similar to replacing essential data with a unique identification symbol to retain its security to make sure the meaning of the data is not compromised. POS tagging serves its applications in the information retrieval system, the natural language parsing system and in machine translation. The following are the major POS classes in Tamil: nouns, verbs, adjectives, adverbs, determiners, post positions, conjunctions, and quantifiers.

7.7.3.4 Stop Word Removal

Stop words are the most commonly used words that are found in any natural language. They do not contribute much to the semantic representation of a sentence. Instead, they are used for a syntactic representation in the sentence formation. In the speech processing task, the stop word removal acts as a preliminary processing phase to enhance the efficiency of recognition and summarization. Table 7.9 gives the sample list of stop words in Tamil language.

TABLE 7.9 List of Sample Stop Words

ஒரு (Oru)
என்று (Eṉṟu)
மற்றும் (Maṟṟum)
இந்த (Inta)
இது (Itu)
என்ற (Eṉṟa)
கொண்டு (Koṇṭu)
என்பது (Eṉpatu)
பல (Pala)
ஆகும் (Ākum)

7.7.3.5 *Structural Feature*

It is used to describe the length of time or the duration of facts provided in a spoken sentence. The structural features include the position of a sentence in the story, speaker type (reporter or not), previous and next speaker type, and the change in speaker type.

7.7.3.6 *Discourse Feature*

The discourse feature is used to show the listener how to interpret what the speaker is saying without affecting the literal meaning (well, oh, like, of course, yeah).

7.7.3.7 *Relevance Feature*

The relevance feature evaluates the suitability of each sentence with its document. The relevance feature will be determined using vector space model score, latent semantic analysis score, and the Markov random walk score.

7.7.4 Related Work on Tamil Text Summarization

Kumar and Devi [63] made use of the graph theoretic scoring technique to assign a score to sentences. Based on that, summary sentences are chosen. To assign a score to the sentences, a term positional and weight-age calculation is inferred in addition to analyzing the frequency of words.

Banu et al. [64] made use of the semantic graph technique to summarize the Tamil documents where the subject, object and predicates are identified from all sentences in the document. Then the summary for the source document is formed by human experts. Here, a triple of subject, object, and predicate semantic normalization is employed to reduce the number of occurrences of nodes in the semantic graph. The triples of subject, object and predicate from the semantic graph is identified by using a leaning technique that is taught using a support vector machine classifier. Then, the summary sentences are extracted from the spoken test documents using the classifier.

Keyan [65] proposed a neural network based multidocument and multilingual (Tamil and English) summarization. Here, individual lines in the document are converted to vector representations, and based on the sentence features, the vector is assigned with a weight score. Then, summarization is performed by selecting summary sentences based on the weight score assigned. In case of multidocument summarization, the summary sentences chosen by a single document summarization module is taken as the input. By making use of similarity and dissimilarity measures, the resultant summary for multidocument is generated. This technique shall be used for both Tamil and English online newspapers summary generation.

7.8 Evaluation Metrics

In the research field, it is necessary to evaluate the performance of summarization results, to benefit the end user by providing a summarizer with better quality. Approaches for evaluating the results of spoken document summarization can be classified as either intrinsic or extrinsic. In case of intrinsic evaluation, the summaries generated by the automatic summary generation system are compared with the summaries generated by human beings manually, and the performance of resultant summaries is analyzed. In case of extrinsic evaluation, the effectiveness of summing up a task or a document is tested and analyzed.

Some aspects of evaluating the summarization results depend on the performer of the evaluation. The summarization evaluation can be performed either manually by human or the evaluation can be done automatically. Evaluations performed by humans can be carried out to match the content of the document, checked for the degree of excellence, grammaticality, usage of less redundant words, and the inclusion of the most important information in the content of the summary. Compared to the human evaluation, the summary that is automatically generated is neutral and it can help minimize human efforts and increase the rate of system development.

7.8.1 ROUGE

Recall-Oriented Understudy for Gisting Evaluation [66] is the commonly employed evaluation metric to analyze the summarization results. It can be in various forms that are discussed below:

7.8.1.1 ROUGE-n

Based on the similarity results of n-grams, a series of 2-grams, 3-grams, and 4-grams is extracted from a summary that is considered as a reference, and is the automatically generated summary. Consider "a" *as* "the number of common n-grams between candidate and reference summary", and "b'" as "the number of n-grams extracted from the reference summary only." The score is calculated using:

$$\mathrm{ROUGE}-\mathrm{n}=\frac{a}{b} \tag{7.3}$$

7.8.1.2 ROUGE-L

This metric makes use of the idea of longest common subsequence (LCS) between the two successions of sentences in the document. The longer the LCS, the two sentences that are chosen for summary generation are said to be more similar. Though this measure is more flexible, it requires all n-grams to be consecutive.

7.8.1.3 ROUGE-SU

This measure is also called as skip bigram and unigram ROUGE. Here, the words can be inserted in between the words of bigrams. Therefore it is not required for the sequence of words to be consecutive.

7.8.2 Precision, Recall, and F-Measure

The summarization systems in general choose the most informative sentences from the spoken document and then generate an extractive summary. The informative sentences that are extracted are simply concatenated together to form a summary, and no changes are made in the original words used in the document. In this scenario, the most commonly used measures such as precision and recall can be used.

In case of precision (P), the sentences that are considered to be the most informative are selected manually by the human and also automatically generated by the system. They are compared with the sentences that are selected automatically by a system. In case of recall (R), the sentences that are considered to be the most informative are selected manually by the human and also automatically generated by the system. They are compared against the human-generated summaries.

$$\mathrm{Recall}=\frac{(\text{number of sentences occuring} \in \text{both system} \wedge \text{ideal summary})}{(\text{number of sentences} \in \text{ideal summary})} \tag{7.4}$$

$$\mathrm{Precision}=\frac{(\text{number of sentences occuring} \in \text{both system} \wedge \text{ideal summary})}{(\text{sentence chosen by the system})} \tag{7.5}$$

F-score is defined as the harmonic average of precision and recall.

$$\text{F}-\text{score}=\frac{2PR}{P+R} \tag{7.6}$$

7.8.3 Word Error Rate

The performance of generating transcripts from the speech waveform can be measured using the word error rate (WER) and is defined as the ratio of number of misclassified words to the total number of words in the spoken content.

$$\text{WER}=\frac{\text{number of words recognized correctly}}{\text{total number of words}} \tag{7.7}$$

7.8.4 Word Recognition Rate

The performance of transcript generation from a spoken data can also be measured using word recognition rate and it is defined as:

$$\text{WRR}=1-\text{WER} \tag{7.8}$$

7.9 Speech Corpora for Tamil Language

A corpus for any language contains a huge volume of structured data. It can be in the form of either written or spoken or in a machine-readable form [67]. In the field of research, there is a great necessity for a language corpus. The speech corpus for other foreign languages are more widely available compared with Indian language. Since Indian languages vary in diversity, accent and dialects, the development of a corpus for Indian language is a time consuming process. However, there are some corpora developed for Indian language and made available for education and research purposes.

The speech data for the database is collected by the joint effort of all the consortium members. The consortium members include IIT Madras, IIIT Hyderabad, IIT Kharagpur, IISc Bangalore, CDAC Mumbai, CDAC Thiruvananthapuram, IIT Guwahati, CDAC Kolkata, SSNCE Chennai, DA-IICT Gujarat, IIT Mandi, and PESIT Bangalore. For speech recording, two voice talents are identified (one male and one female) for each language. Text in each language is identified for reading and is read in an anechoic chamber. A total of 40h of speech data is collected for a language—20h of native (mono) data (10h each of male and female data) and 20h of English data recorded by these native speakers (10h each of male and female data).

The Central Institute of Indian Language (CIIL) corpus is a collection of a vast amount of text documents in Tamil language. It contains 2.6 million words that are based on various domains such as cooking tips, news articles, biographies, and agriculture.

The Linguistic Data Consortium (LDC) corpus includes a collection of news text that is recorded in a noisy environment and a stereo recording is used to record the news read by different people. The news text is read by a group of people, and they are categorized based on age, gender, and a different environment. The entire speech file is saved as a .ZIP file, and it includes the corresponding transcripts of speech data labeled at sentence level.

The Indian Language Technology Proliferation and Technology Centre corpus contains more than 62,000 audio files in Tamil language of 1000 speakers. The dataset size is around 5.7 GB and the content is prepared for the agricultural domain. The dataset collection includes a .doc file with the list of words and their corresponding phonetic representations, along with the transcripts for each audio file.

7.10 Conclusion

Although speech-to-text conversion (STT) machines aim at providing benefits for the deaf or people who can't speak, it is difficult to review, retrieve and reuse speech transcripts. Hence, when the speech to text conversion module is combined with the summarization, the applications further increase in educational fields as well. This chapter discussed the need for speech summarization, various issues in the summarization of a spoken document, supervised, and unsupervised summarization algorithms. Isolated Tamil speech recognition was performed using a sample set of Tamil spoken words. In addition, state-of-the-art recognition techniques were used, and analysis was performed. Also, the summarization of speech data in Tamil language is explored, along with related work on text summarization. The features used in the summarization of a spoken document are analyzed and compared, based on the various forms of input into the spoken document.

References

[1] K. Zechner, Summarization of Spoken Language—Challenges, Methods, and Prospects, Language Technologies Institute Carnegie Mellon University, 2002.

[2] B. Chen, S.-H. Lin, Y.-M. Chang, J.-W. Liu, Extractive speech summarization using evaluation metric-related training criteria, Inf. Process. Manag. 49 (2013) 1–12.

[3] B. Chen, S.-H. Lin, A risk-aware modeling framework for speech summarization, IEEE Trans. Audio Speech Lang. Process. 20 (1) (2012) 199–210.

[4] D. Shen, J.-T. Sun, H. Li, Q. Yang, Z. Chen, Document summarization using conditional random fields, in: Proceedings of the 20th International Joint Conference on Artificial Intelligence, 2007, pp. 2862–2867.

[5] M.A. Fattah, F. Ren, GA, MR, FFNN, PNN and GMM based models for automatic text summarization, Comput. Speech Lang. 23 (1) (2009) 126–144.

[6] J. Kupiec, et al., A trainable document summarizer, in: Proc. of the Annual International ACM SIGIR Conference, 1995, pp. 68–73.

[7] A. Kolcz, et al., Summarization as feature selection for text categorization, in: Proc. ACM Conference on Information and Knowledge Management, 2001, pp. 365–370.
[8] M. Galley, Skip-chain conditional random field for ranking meeting utterances by importance, in: Proc. Empirical Methods in Natural Language Processing, 2006, pp. 364–372.
[9] C.-I. Tsai, H.-T. Hung, K.-Y. Chen, B. Chen, Extractive speech summarization leveraging convolutional neural network techniques, in: GlobalSIP IEEE, 2016.
[10] S.-H. Liu, K.-Y. Chen, B. Chen, E.-E. Jan, H.-M. Wang, H.-C. Yen, W.-L. Hsu, A margin-based discriminative modeling approach for extractive speech summarization, in: Proc. of APSIPA ASC, 2014.
[11] L.S. Lee, B. Chen, Spoken document understanding and organization, IEEE Signal Process. Mag. 22 (5) (2005) 42–60.
[12] B. Chen, Y.-T. Chen, Extractive spoken document summarization for information retrieval, Pattern Recogn. Lett. 29 (4) (2008) 426–437.
[13] Y. Gong, X. Liu, Generic text summarization using relevance measure and latent semantic analysis, in: Proc. ACM SIGIR Conf. R&D Inf. Retrieval, 2001, pp. 19–25.
[14] M. Hirohata, Y. Shinnaka, K. Iwano, S. Furui, Sentence-extractive automatic speech summarization and evaluation techniques, Speech Commun. 48 (9) (2006) 1151–1161.
[15] G. Murray, S. Renals, J. Carletta, Extractive summarization of meeting recordings, in: Proc. Eur. Conf. Speech Commun. Technol, 2005, pp. 593–596.
[16] J. Carbonell, J. Goldstein, The use of MMR, diversity-based reranking for reordering documents and producing summaries, SIGIR (1998).
[17] G. Erkan, D.R. Radev, LexRank: graph-based lexical centrality as salience in text summarization, J. Artif. Intell. Res. 22 (2004) 457–479.
[18] R. Mihalcea, P. Tarau, TextRank: bringing order into texts, in: Proc. Conference on Empirical Methods in Natural Language Processing, 2005, pp. 404–411.
[19] X. Wan, J. Yang, Multi-document summarization using cluster-based link analysis, in: Proc. the Annual International ACM SIGIR Conference on Research and Development in Information Retrieval, 2008, pp. 299–306.
[20] H. Daumé III, D. Marcu, Bayesian query focused summarization, in: Proc. Annual Meeting of the Association for Computational Linguistics, 2006, pp. 305–312.
[21] Y.T. Chen, B. Chen, H.M. Wang, A probabilistic generative framework for extractive broadcast news speech summarization, IEEE Trans. Audio Speech Lang. Process. 17 (2009) 95–106.
[22] A. Khan, N. Salim, A review on abstractive summarization methods, J. Theor. Appl. Inf. Technol. 59 (1) (2014).
[23] P.E. Genest, G. Lapalme, Framework for abstractive summarization using text-to-text generation, in: Proceedings of the Workshop on Monolingual Text-To-Text Generation, 2011, pp. 64–73.
[24] R. Barzilay, et al., Information fusion in the context of multi-document summarization, in: Proceedings of the 37th annual meeting of the Association for Computational Linguistics on Computational Linguistics, 1999, pp. 550–557.
[25] R. Barzilay, K.R. McKeown, Sentence fusion for multidocument news summarization, Comput. Linguist. 31 (2005) 297–328.
[26] S.M. Harabagiu, F. Lacatusu, Generating single and multi-document summaries with gistexter, in: Document Understanding Conferences, 2002.
[27] C.-S. Lee, et al., A fuzzy ontology and its application to news summarization, Trans. Syst. Man Cybern. Part B: Cybern. IEEE 35 (2005) 859–880.

[28] H. Tanaka, et al., Syntax-driven sentence revision for broadcast news summarization, in: Proceedings of the 2009 Workshop on Language Generation and Summarisation, 2009, pp. 39–47.
[29] P.-E. Genest, G. Lapalme, Fully abstractive approach to guided summarization, in: Proceedings of the 50th Annual Meeting of the Association for Computational Linguistics: Short Papers, vol. 2, 2012, pp. 354–358.
[30] H. Saggion, G. Lapalme, Generating indicative-informative summaries with sumUM, Computational Linguistics 28 (2002) 497–526.
[31] C.F. Greenbacker, Towards a framework for abstractive summarization of multimodal documents, ACL HLT (2011) 75.
[32] P.E. Genest, G. Lapalme, Framework for abstractive summarization using textto-text generation, in: Proceedings of the Workshop on Monolingual Text-To-Text Generation, 2011, pp. 64–73.
[33] I.F. Moawad, M. Aref, Semantic graph reduction approach for abstractive text summarization, in: 2012 Seventh International Conference on Computer Engineering & Systems (ICCES), 2012, pp. 132–138.
[34] I. Mahadevan, Early Tamil Epigraphy From the Earliest Times to the Sixth Century A.D., Harvard Oriental Series, vol. 62, Harvard University Press, Cambridge, 2003. ISBN 0-674-01227-5.
[35] S.B. Steever, W. Bright, P.T. Daniels, Tamil writing, in: The World's Writing Systems, Oxford University Press, New York, 1996, pp. 426–430. ISBN 0-19-507993-0.
[36] R. Thagarajan, A.M. Natarajan, M. Selvam, Syllable modeling in continuous speech recognition for Tamil language, Int. J. Speech Technol. 12 (2009) 47–57.
[37] S. Karpagavalli, E. Chandra, Phoneme and word based model for Tamil speech recognition using GMM-HMM, in: International Conference on Advanced Computing and Communication Systems (ICACCS -2015), 2015.
[38] C. Vimala, V. Radha, Speaker independent isolated speech recognition system for Tamil language using HMM, Proc. Eng. 30 (2012) 1097–1102.
[39] C. Vimala, V. Radha, Suitable feature extraction and speech recognition technique for isolated Tamil spoken words, Int. J. Comput. Sci. Inf. Technol. 5 (1) (2014) 378–383.
[40] C. Vimala, V. Radha, Isolated speech recognition system for Tamil language using statistical pattern matching and machine learning techniques, J. Eng. Sci. Technol. 10 (5) (2015) 617–632.
[41] A. Lakshmi, H.A. Murthy, A new approach to continuous speech recognition in Indian languages, in: Proc. National Conference on Communication, Mumbai, India, 2008, pp. 277–281.
[42] V. Radha, M. Krishnaveni, Isolated word recognition system for Tamil spoken language using Back propagation neural network based OnLpcc features, Int. J. (CSEIJ) 1 (4) (2011).
[43] A. Graves, A.-R. Mohamed, G. Hinton, Speech recognition with deep recurrent neural networks. in: ICASSP, IEEE International Conference on Acoustics, Speech and Signal Processing—Proceedings, vol. 38, 2013. https://doi.org/10.1109/ICASSP.2013.6638947.
[44] M. Gales, S. Young, The application of hidden Markov models in speech recognition, Found. Trendsin Signal Process. 1 (3) (2007).
[45] C. Dalmiya, V. Dharun, K. Rajesh, An efficient method for Tamil speech recognition using MFCC and DTW, in: IEEE Conference on Information and Communication Technologies (ICT), 2013, pp. 1263–1268.
[46] V. Kamakshi Prasad, T. Nagarajan, H.A. Murthy, Continuous speech recognition using automatically segmented data at syllabic units, in: Proceedings of the Sixth International Conference on Signal Processing, ICSP, 2002, pp. 235–238.
[47] Q. Li, Y. Huang, An auditory-based feature extraction algorithm for robust speaker identification under mismatched conditions, IEEE Trans. Audio Speech Lang. Process. 19 (6) (2011) 1791–1801.

[48] R. Schluter, L. Bezrukov, H. Wagner, H. Ney, Gammatone features and feature combination for large vocabulary speech recognition, in: IEEE International Conference on Acoustics, Speech and Signal Processing, ICASSP, vol. 4, 2007. pp. IV-649–IV-652.
[49] S. Sundarapandiyan, N. Shanthi, M. Mohamed Yoonus, Syllable based Tamil language continuous robust speech recognition using MGDFGWCC with DNN-HMM, IJCTA 9 (7) (2016) 3391–3400.
[50] S. Maskey, J. Hirschberg, Comparing Lexical, Acoustic/Prosodic, Structural and Discourse Features for Speech Summarization, Columbia University, New York, 2005. Interspeech.
[51] J. Zhang, H. Chan, P. Fung, L. Cao, A comparative study on speech summarization of broadcast news and lecture speech, in: Proceedings of the Interspeech, ISCA Archive, Grenoble, France, 2007, pp. 2781–2784.
[52] V. Dhanalakshmi, M. Anandkumar, G. Shivapratap, K.P. Soman, S. Rajendran, Tamil POS tagging using linear programming, Int. J. Recent Trends Eng. 1 (2) (2009) 166–169.
[53] H. Christensen, Y. Gotoh, B. Kolluru, S. Renalset, Are extractive text summarisation techniques portable to broadcast news? in: Proceedings of Automatic Speech Recognition and Understanding Workshop, IEEE Press, New York, NY, 2003, pp. 489–494.
[54] S. Maskey, J. Hirschberg, Comparing lexical, acoustic/prosodic, structural and discourse features for speech summarization, in: Proceedings of Interspeech, ISCA Archive, Grenoble, France, 2005, pp. 621–624.
[55] S. Maskey, J. Hirschberg, Summarizing speech without text using Hidden Markov Models, in: Proceedings of the Human Language Technology Conference of the NAACL (Companion Volume: Short Papers), Association for Computational Linguistics, Stroudsburg, PA, 2006, pp. 89–92.
[56] J. Zhang, H. Chan, P. Fung, Improving lecture speech summarization using rhetorical information, in: Proceedings of the IEEE Workshop on Automatic Speech Recognition and Understanding, IEEE Press, New York, NY, 2007, pp. 195–200.
[57] S. Xie, D. Hakkani-Tur, B. Favre, Y. Liu, Integrating prosodic features in extractive meeting summarization, in: Proceedings of the 11th Biannual IEEE Workshop on Automatic Speech Recognition and Understanding, IEEE Press, New York, NY, 2009, pp. 387–391.
[58] Z. Zhang, P. Fung, Active learning with semi-automatic annotation for extractive speech summarization, ACM Trans. Speech Lang. Process. 8 (4) (2012) 1–25.
[59] J. Zhang, H. Yuan, Speech summarization without lexical features for Mandarin presentation speech, in: International Conference on Asian Language Processing, IEEE, 2013.
[60] S.-H. Liu, K.-Y. Chen, B. Chen, E.-E. Jan, H.-M. Wang, H.-C. Yen, W.-L. Hsu, A margin-based discriminative modeling approach for extractive speech summarization, in: Proc. of APSIPA ASC, 2014.
[61] T. Hasan, M. Abdelwahab, S. Parthasarathy, C. Busso, Y. Liu, Automatic composition of broadcast news summaries using rank classifiers trained with acoustic and lexical features, in: ICASSP IEEE, 2016.
[62] S. Stephanie, C. Walker, H. Lee, RT-03 MDE Training Data Speech LDC2004S08 [Online], https://catalog.ldc.upenn.edu/LDC2004S08, 2004.
[63] S. Kumar, V.S. Ram, S.L. Devi, Text extraction for an agglutinative language, in: Proc. J. Lang. India, 2011, pp. 56–59.
[64] M. Banu, C. Karthika, P. Sudarmani, T.V. Geetha, Tamil document summarization using semantic graph method, in: Proceedings of International Conference on Computational Intelligence and Multimedia Applications, 2007, pp. 128–134.
[65] M.K. Keyan, K.G. Srinivasagan, Multi-document and multi-lingual summarization using neural networks, in: Proceedings of International Conference on Recent Trends in Computational Methods, Communication and Controls, 2012, pp. 11–14.

[66] C.-Y. Lin, Rouge: a package for automatic evaluation of summaries, in: Text Summarization Branches Out: Proceedings of the ACL-04 Workshop, 2004, pp. 74–81.
[67] A. Baby, A. Leela Thomas, N.L. Nishanthi, TTS Consortium, Resources for Indian languages, in: Computer Science and Engineering, IIT Madras, 2016, pp. 37–43.

Further Reading

N. Dey, A.S. Ashour, W.S. Mohamed, N. Nhu, Introduction: studies in speech signal processing, natural language understanding, and machine learning. in: Acoustic Sensors for Biomedical Applications, 2019, pp. 1–5, https://doi.org/10.1007/978-3-319-92225-6_1.
N. Dey, A.S. Ashour, Direction of Arrival Estimation and Localization of Multi-Speech Sources, Springer International Publishing, 2018.

Chapter 8

Classifying Recurrent Dynamics on Emotional Speech Signals

Sudhangshu Sarkar* **and Anilesh Dey**[†]
[*]*Department of Electrical Engineering, Narula Institute of Technology, Kolkata, India,*
[†]*Department of Electronics and Communication Engineering, Narula Institute of Technology, Kolkata, India*

8.1 Introduction

Speech signal processing is a vast field of study, which contributes highly to human computer interaction (HCI) studies. Speech is simply the best carrier of information in human communication systems. It contains a lot of information apart from the verbal message, that is, speaker identification, speaker's age, sex, locality, emotion, etc. Recognition of emotion contained in a speech signal is one of the fastest growing areas of interest in HCI study. Emotion in a speech plays an important role in expressing feelings. Based on different emotions, human speak in different ways, and the characteristics of speech changes, accordingly. Humans don't need practice to recognize the emotional state of a speaker; it comes naturally [1]. However, it is a complex process when it's implemented in a machine.

In this regard, a literary review of the past studies on speech-based emotion recognition systems was made. Researchers have proposed many techniques for emotion-based speech recognition [2, 3]. Schuller et al. [4] reported the continuous use of hidden Markov models (HMM) for speech-emotion recognition. The same group further extended their work in 2004 by combining acoustic features with linguistic data for a healthy emotion detection using support vector machine (SVM) [5]. Lin et al. [6] used the aforementioned two methods, namely HMM and SVM for classification of five dissimilar states of emotion, that is, annoyance, pleasure, sorrow, shock, and impartial emotion. Lalitha et al. [7] reported the use of time-domain speech features like pitch and prosody for recognition of seven different emotional states. Kamal et al. [8] predicted protein structures from images using HMM. Chapman Kolmogrov. Dey et al. [9,10] analyzed the progressiveness of acoustic waves in biomedical technology.

Intelligent Speech Signal Processing. https://doi.org/10.1016/B978-0-12-818130-0.00008-8

Identifying suitable features that characterize different emotions is an important process for developing a speech emotion recognition (SER) system [11]. Altrov et al. [12] identified the power of verbal communication and civilization on the accepting of language emotions. An effective discussion mainly depends on how we communicate our own emotions, how we recognize those of others, and how sufficient our reaction is to their emotions. Cowie et al. [13,14] illustrated that emotions have an essential role in our lives since they are typically present in everyday communication. Park and Sim [15] showed emotion detection by DRNN. Their paper found that pitch was a significant component in the identification of emotion. Therefore the basic accurate detection acoustical features [16,17] were analyzed for speech sound with emotion. The study of event-related potentials (ERPs) is recognized as a useful technique for exploring intelligent mechanisms of processing emotional speech [18]. Tao et al. [19] attempts to create exciting dialogue via “strong,” “average,” and “weak” classifications using various models like a linear modification model (LMM), a Gaussian mixture model (GMM), a classification and regression tree model (CART). Kang and Li [19] analyzed neutral-emotional speech by using prosody conversion. In 2010, Wu et al. [20] presented an advance to hierarchical prosody translation for an exciting language mixture. Jia et al. [21] adopted “Emotional Audio Visual Speech Synthesis Based on PAD,” while Dey and Ashour [22, 23] discussed arrival estimation of localized multispeech sources. An emotional text-to-speech system [24] is required for emotion-based speech recognition. Neural networks [25–27] have exhibited remarkable success to link the responsive space in communication signals. The direction of speech resources on a localized level has been eminently described by Dey and Ashour [28–30]. least squares regression [31] is one of the noted methods for speech emotion recognition. An ideal scientific SER system would be one that can develop real life and loud talking to recognize different state of emotions. In this paper, we have attempted to classify the recurrent dynamics of two different emotions, namely anger and normal, with the help of recurrence plot, phase space plot, and recurrence based parameters. The investigation was performed in both noise-free and on noisy environment to establish the suitability of the proposed method.

8.2 Data Collection and Processing

A healthy male volunteer (age 23 years old) was asked to participate in the study. He was informed about the details of the study, and a written consent to participate was obtained. Two types of speech signals were acquired from the volunteer using microphone Behringer C-1U, when he uttered eight different sentences (in the Bengali language) in angry and normal emotion. In order to process the speech signals, the sampling frequency was taken as 16 KHz in using Audacity version 1.3.6. in the Electronics and Communication Engineering Department, Narula Institute of Technology, Agarpara, Kolkata.

8.3 Research Methodology

Phase space approach was used to investigate the nonlinear properties of the speech signals. A phase space was reconstructed for each speech signal with appropriate time delay and proper embedding measurement.

8.3.1 Phase Space Reconstruction

The condition of a dynamical method is able to be illustrated in a space called phase space. A phase space is a multidimensional space, in which every point correlates with one state of the dynamical system [32]. The path traced by the phase space diagram of a system over time describes its evolution from an initial state. This is known as phase space trajectory.

The basic problem is that the information about all the variables governing the system is usually not obtained from its time series. Most of the time the series is single valued. Although numerous concurrent measurements can be performed; they may not coat each degree of freedom of the arrangement. Though the use of the time-delay embedding theorem [33] allows the recovery of lost information, and it becomes possible to construct the phase space diagram of a scheme from its period sequence. This method of phase space reconstruction requires the determination of the principles of suitable time delay τ and proper embedding measurement m.

Determining the most favorable value of time delay τ for the phase space reformation, for any time series $\{x(t)\}_{t=1}{}^{N}$ at a given state $x(t)$, τ is one of the proper values of time delay, which divulges utmost novel information through dimension at $x(t+\tau)$. The auto mutual information (AMI) technique [34] is usually adapted the proper value of τ. The AMI of a time series for a given τ is calculated using Eq. (8.1) [35]. The optimal value of τ is that one for which AMI(τ) reaches its first minimum [34].

$$\mathrm{AMI}(\tau) = \sum_{t=1}^{N-\tau} P[x(t), x(t+\tau)] \log \left(\frac{P[x(t), x(t+\tau)]}{P[x(t)]P[x(t+\tau)]} \right) \tag{8.1}$$

where $\tau = [1, 2, \ldots, N-1]$ and $P[\,]$ denotes the probability.

Embedding dimension is a measure of the least element of the phase space of the reconstructed characteristic of a dynamical system [36, 37]. Kennel et al. [38] have anticipated the method of false nearest neighbor (FNN) to determine the minimum satisfactory embedding dimension m. The FNN algorithm can be described as follows.

For every point $\vec{R_i}$ in the time series, its adjacent neighbor $\vec{R_j}$ is searched in an m- dimensional space. The space $\left\| \vec{R_i} - \vec{R_j} \right\|$ is calculated. Both the points are iterated and $\vec{R_i}$ is computed as given in Eq. (8.2).

$$R_i = \frac{|R_{i+1} - R_{j+1}|}{\left\| \vec{R_i} - \vec{R_j} \right\|} \tag{8.2}$$

If the computed value of $\vec{R_i}$ goes beyond a specified heuristic verge $\vec{R_t}$, this point is regarded as having a fake nearby neighbor. The minimal embedding length is obtained when the percentage of FNN at a given dimension reaches zero.

8.3.2 Recurrence Plot Analysis

Recurrence plot (RP) is one of the efficient graphical methods designed to find the hidden nonlinear structure of the phase spaces, introduced by Eckmann et al. [39]. For any two points x_i, x_j in a phase space, the distance among x_i and x_j can be calculated by $\|x_i - x_j\|$. Then, the recurrence between two points x_i and x_j is given by.

$$R_{ij} = \Theta\left(\varepsilon - \|x_i - x_j\|\right) \tag{8.3}$$

where Θ represents Heaviside function.

From the definition, it follows that the entries in the matrix $(R_{ij})_{N\times N}$ (N being the span of the trajectory of the phase space) are either 1 or 0. The number "1" is represented by a black dot. On the other hand, "0" is represented by a white dot. So, an RP is a visual representation of a phase space by two colors. From the structure of the RP, various dynamical patterns of a complex dynamic can be described, such as periodicity, quasiperiodicity, noise effect, nonstationary behavior, and a chaotic nature. It indicates that classification between two different dynamics can be made by RP analysis. Fig. 8.1(A) and (B) show the recurrence plot of a speech signal in angry emotion and normal emotion, respectively.

If two points x_i and x_j are recurrent, we say that there is an isometry. Two points x_i and x_j in a phase space are said to be in consecutive isometry if.

$$R_{ij} = \Theta\left(\varepsilon - \|x_{i+L} - x_{j+L}\|\right) \tag{8.4}$$

where$L \in Z^+$.

Since periodicity and aperiodicity of a phase space are the reflection of isometry, so the nature of the dynamics can be described by it. In fact, complexity of the RP decreases as the consecutive isometry increases [40, 41].

8.4 Numerical Experiments and Results

In this work, speech signals in angry and normal emotion were analyzed using RP analysis for two cases, that is, noise free and noisy conditions. Two informative parameters, namely isometry and consecutive isometry were calculated for each case. For both cases, proper time delay and embedding dimension were

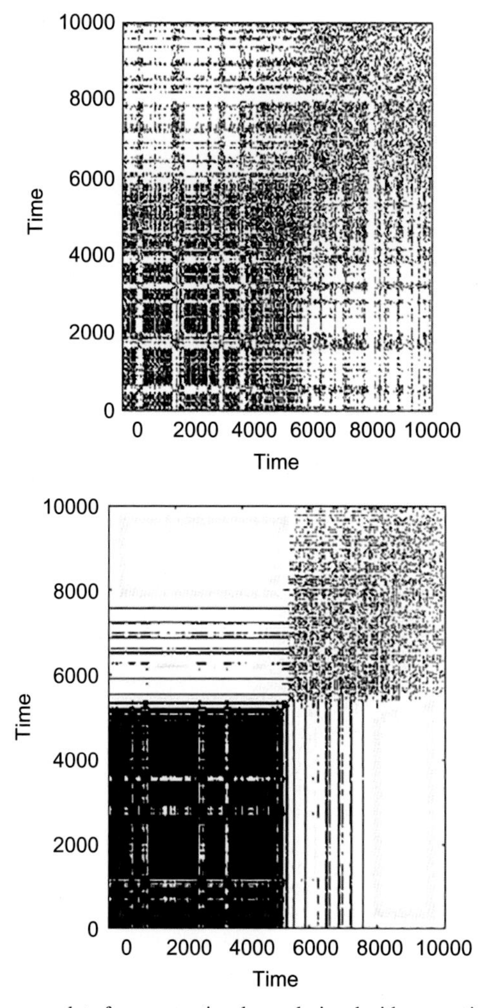

FIG. 8.1 (A) Recurrence plot of angry emotional speech signal with proper time delay and embedding dimension. (B) Recurrence plot of normal emotional speech signal with proper time delay and embedding dimension.

recalculated. Fig. 8.2(A) shows the time delay for a speech signal in both the angry and normal emotion, which are represented by red and blue lines respectively. In order to calculate the probability, 17 bins were considered.

From the figure, it is observed that the AMI was minimum at $\tau = 60$ in the case of normal speech, whereas it was minimum at $\tau = 43$ for the angry speech signal. It suggested that the optimal time delay was different for the same speech in two different emotions. In fact, it was higher in the case of normal speech than that of the speech in angry emotion.

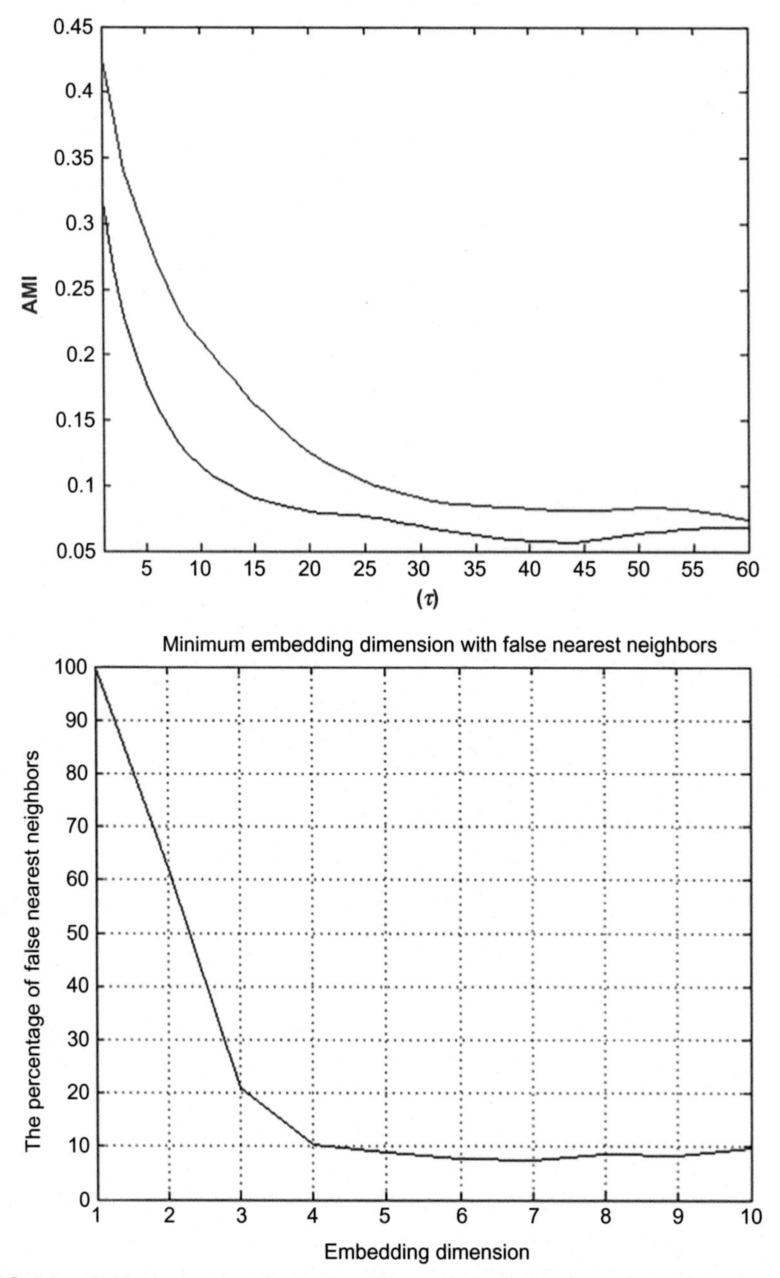

FIG. 8.2 (A) Fluctuation of AMI with time delay $\tau \in [1, 60]$. (B) Fluctuation of FNN for a speech signal in the angry emotion with embedding. (C) A phase space diagram for speech signals in two different emotional conditions.

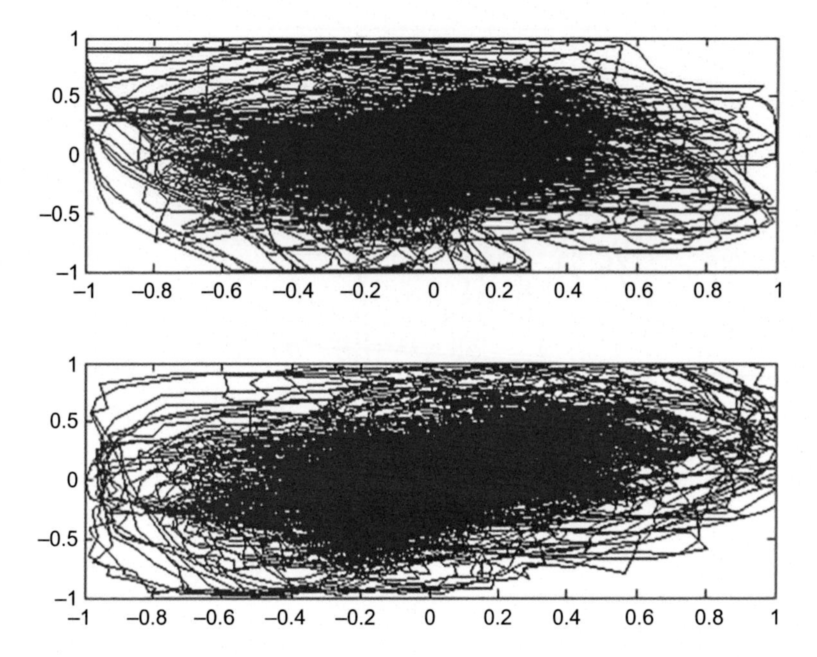

FIG. 8.2—cont'd

The percentage of FNN was calculated for embedding dimensions ranging from 1 to 10 with variable time delay τ varies from 1 to 60. Fig. 8.2(B) shows the fluctuation of FNN with embedding dimension for a speech signal in the angry emotion. The dimension, for which the percentage of FNN was minimum, was taken as the suitable embedding dimension.

Dimension ranging from 1 to 10 with $\tau=43$. The phase space diagram was reconstructed for the speech signal for angry and normal emotion with proper time delay [42, 43] for each case as given in Fig. 8.2(C).

From Fig. 8.2(C), it can be seen that the number of outliers increased in the angry emotion compared to the normal emotion. It also can be seen that the normal emotion speech signal exhibited denser orbit than the angry emotion speech signal. These results suggest that different dynamics benefit different emotional states.

8.4.1 Noise-Free Environment

As stated above, isometry is the measure of recurrence in any phase space. It is calculated as

$$\text{Isometry} = \frac{1}{N^2}\sum_{i,j=1}^{N} R_{i,j}(\varepsilon) \tag{8.5}$$

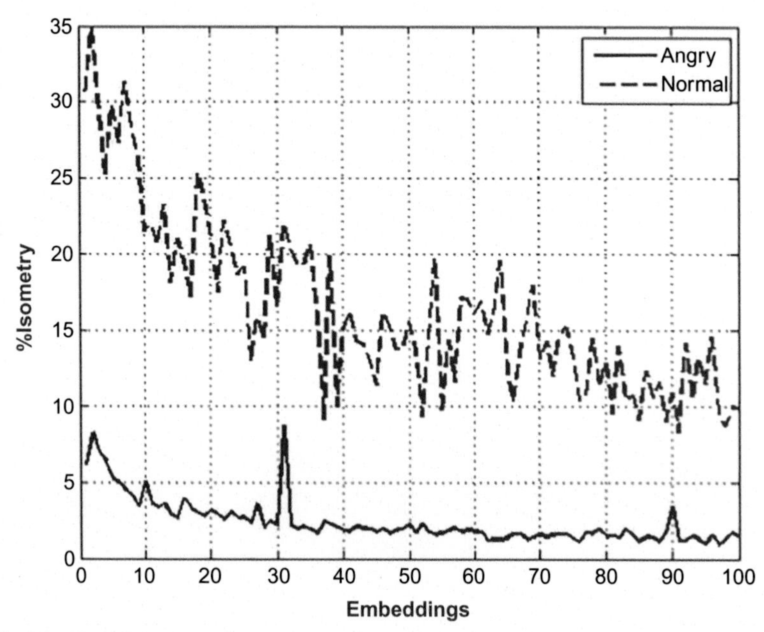

FIG. 8.3 Embedding plot of %isometry with proper time delay for angry and normal emotional speech in a noise-free environment.

where N is the number of speech samples and $R_{i,\ j}(\varepsilon)$ is the repetition matrix equivalent to a threshold of ε. Isometry is represented as the quantity of isometric recurrences articulated as a fraction of the entire quantity of pairs of vectors contrast in the sample (i.e., %isometry) [44]. The design of %isometry or any of its copied frameworks as a function of embedding dimension is considered as an embedding plot. The embedding plot of %isometry, represented in Fig. 8.3, shows the comparison between the %isometry of two speech signals in angry and normal emotion in a noise-free environment.

Fig. 8.3 clearly shows that the %isometry of the normal emotional speech signal was higher than that of the angry speech signal. It suggests that the autocorrelation was higher in the case of the normal speech signal. The same observation is also made for %consecutive isometry, represented in Fig. 8.4. (See Fig. 8.5.)

8.4.2 Noisy Environment

To investigate the effect of noise on the isometry, a Gaussian noise was added, given by $\varphi(\xi) = e^{\frac{-\xi^2}{2}}$, where $\varphi(\xi)$ is a Gaussian random variable.

When the energy of the signal is strong in the region of a restricted time interim (especially if its total energy is limited), one may calculate the energy spectral compactness. However, more commonly used is the power spectrum.

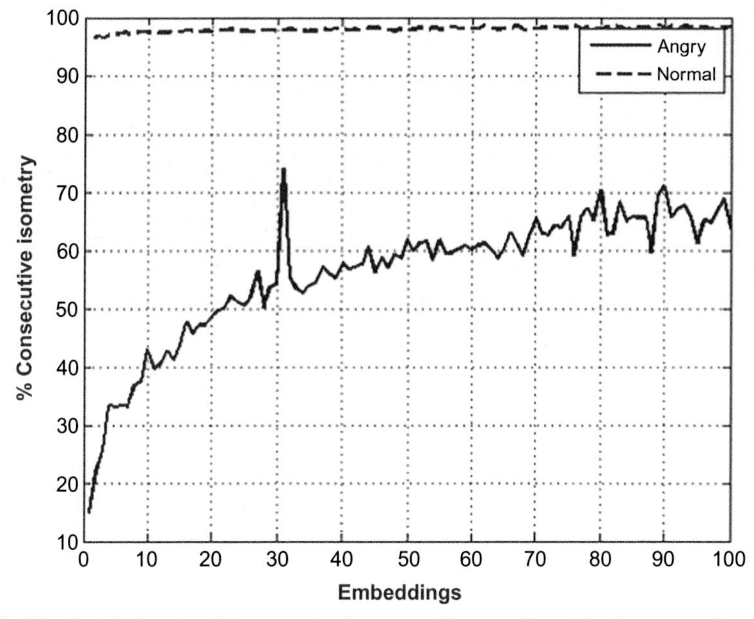

FIG. 8.4 Embedding plot of %consecutive isometry with proper time delay for angry and normal emotional speech in a noise-free environment.

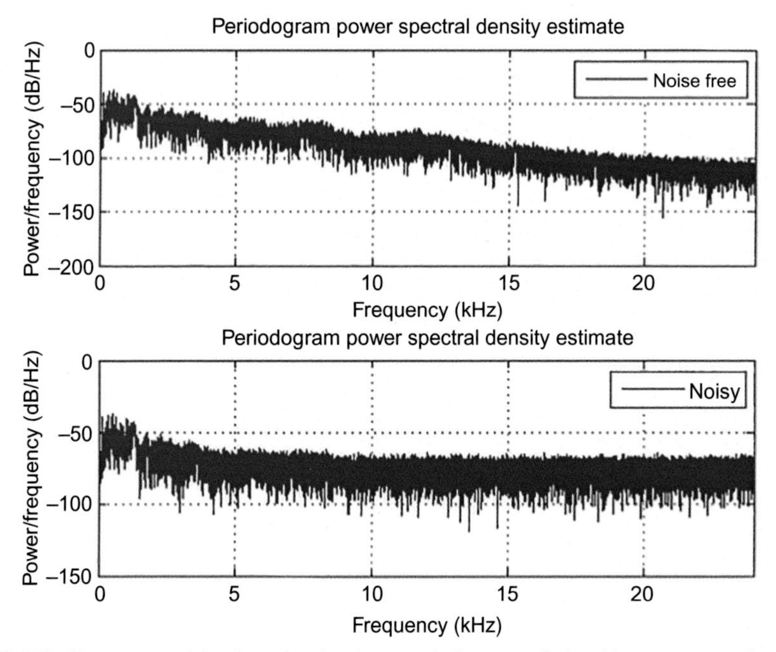

FIG. 8.5 Power spectral density estimation (spectrum) of one speech signal in an angry emotion in a noise-free and noisy environment.

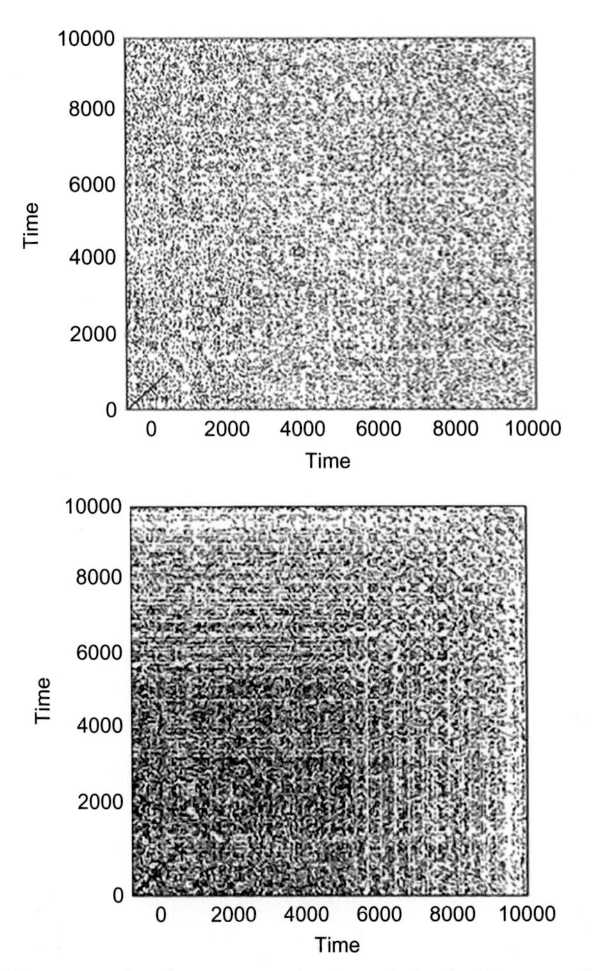

FIG. 8.6 (A) Recurrence plot of an angry emotional speech signal with s proper time delay and embedding dimension after the addition of noise. (B) Recurrence plot of a normal emotional speech signal with a proper time delay and embedding dimension after the addition of noise.

In the experiment, it was noticed that the power frequency of the signal became higher in the power spectrum after the addition of noise.

Fig. 8.6(A) and (B) shows the recurrence plot of a speech signal in angry emotion and usual emotion, respectively. After the addition of noise.

In a noisy condition, it was observed that the recurrence plots were more complex with lots of dark dots. It suggested that the addition of noise in the signal increased the recurrence rate.

Both %isometry and %consecutive isometry were again calculated for the speech signals in angry and normal emotions after adding the noise. Figs. 8.7 and 8.8 show the embedding plots of %isometry and %consecutive isometry,

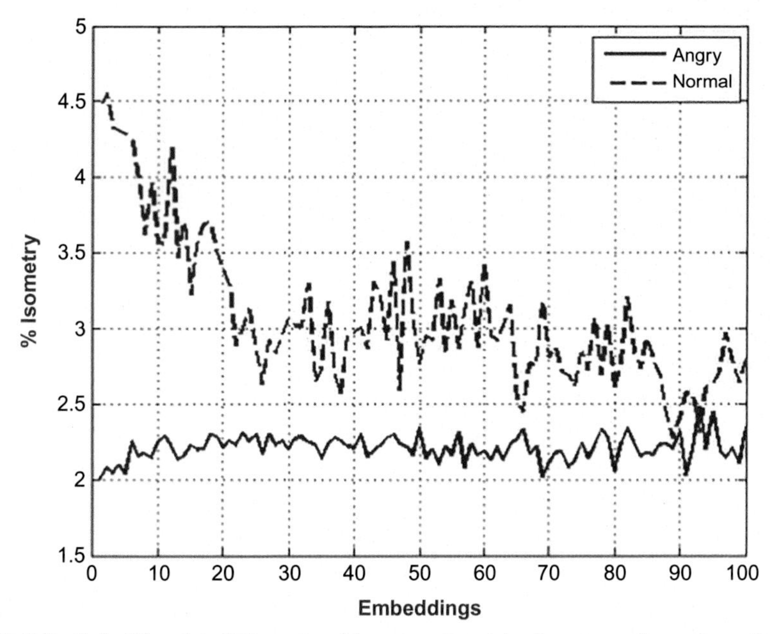

FIG. 8.7 Embedding plot of %isometry with a proper time delay for angry and normal emotional speech.

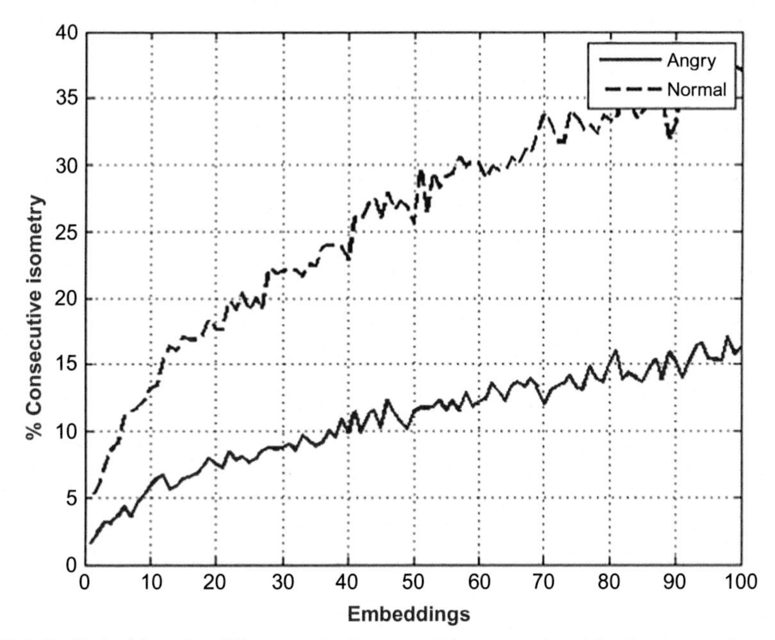

FIG. 8.8 Embedding plot of %consecutive isometry with a proper time delay for angry and normal emotional speech.

for speech signals in angry and normal emotion, respectively, after adding noise. It is evident from the figures that the speech signal in normal emotion exhibited more %isometry and %consecutive isometry than the speech signal in an angry emotion, similar to that of a noise-free condition. These results indicated the robustness of the two parameters even in the noisy condition.

8.5 Conclusion

Speech signals were acquired from a healthy male volunteer at a sampling rate of 16 KHz, when the volunteer uttered eight sentences in the Bengali language in normal and angry emotion. Phase space analysis and recurrence analysis were performed on the speech signals to classify normal and angry emotion conditions. During the editing of the phase space diagram, the proper values of time delay and the embedding dimension were approximated using auto mutual information (AMI) and false nearest neighbor (FNN) mechanism, respectively. The proper time delay was found to be more for normal emotion than angry emotion. The phase space plot for normal emotion had a denser orbit and lower number of outliers than that of the angry emotion. The recurrence plots were drawn and correlated parameters, specifically %isometry and %consecutive isometry, were calculated for the speech signals acquired during normal and angry emotion in noise-free and noisy environments. In both noise-free and noisy environments, %isometry and %consecutive isometry values were greater for normal emotion than angry emotion. This suggested that these parameters are robust to environmental noise and can be used to develop a SER system for classifying different emotional states.

References

[1] N.J. Gogoi, J. Kalita, Emotion recognition from acted Assamese speech, Int. J. Innov. Res. Sci. Eng. Technol. 4 (6) (2015).

[2] D. Ververidis, C. Kotropoulos, Emotional speech recognition: resources, features, and methods, Speech Comm. 48 (2006) 1162–1181.

[3] M. El Ayadi, et al., Survey on speech emotion recognition: features, classification schemes, and databases, Pattern Recogn. 44 (2011) 572–587.

[4] B. Schuller, G. Rigoll, M. Lang, Hidden Markov model-based speech emotion recognition, in: 2003 International Conference on Multimedia and Expo. ICME'03. Proceedings (Cat. No.03TH8698), Baltimore, MD, USA, 2003, pp. 1–401.

[5] B. Schuller, G. Rigoll, M. Lang, Speech emotion recognition combining acoustic features and linguistic information in a hybrid support vector machine-belief network architecture, in: 2004 IEEE International Conference on Acoustics, Speech, and Signal Processing, Montreal, Que, 2004, pp. 1–577.

[6] Y.-L. Lin, G. Wei, Speech emotion recognition based on HMM and SVM, in: 2005 International Conference on Machine Learning and Cybernetics, Guangzhou, China, vol. 8, 2005, pp. 4898–4901.

[7] S. Lalitha, et al., Speech emotion recognition, in: 2014 International Conference on Advances in Electronics, Computers and Communications (ICAECC), 2014, pp. 1–4.
[8] M.S. Kamal, L. Chowdhury, M.I. Khan, A.S. Ashour, J.M.R. Tavares, N. Dey, Hidden Markov model and Chapman Kolmogrov for protein structures prediction from images, Comput. Biol. Chem. 68 (2017) 231–244.
[9] N. Dey, A.S. Ashour, W.S. Mohamed, N.G. Nguyen, Acoustic sensors in biomedical applications, in: Acoustic Sensors for Biomedical Applications, Springer, Cham, 2019, pp. 43–47.
[10] N. Dey, A.S. Ashour, W.S. Mohamed, N.G. Nguyen, Acoustic wave technology, in: Acoustic Sensors for Biomedical Applications, Springer, Cham, 2019, pp. 21–31.
[11] S.G. Koolagudi, et al., IITKGP-SESC: speech database for emotion analysis, in: International Conference on Contemporary Computing, 2009, pp. 485–492.
[12] R. Altrov, H. Pajupuu, The influence of language and culture on the understanding of vocal emotions, J. Estonian Finno-Ugric Linguistics 6 (3) (2015).
[13] R. Cowie, N. Sussman, A. Ben-Ze'ev, Emotions: concepts and definitions, in: P. Petta, C. Pelachaud, R. Cowie (Eds.), Emotionoriented Systems: The HUMAINE Handbook, Springer, Berlin, Heidelberg, 2011, pp. 9–31.
[14] E. Douglas-Cowie, N. Campbell, R. Cowie, P. Roach, Emotional speech: towards a new generation of databases, Speech Comm. 40 (1) (2003) 33–60.
[15] C.-H. Park, K.-B. Sim, Emotion recognition and acoustic analysis from speech signal, in: Proceedings of the International Joint Conference on Neural Networks, 2003, Portland, OR, vol. 4, 2003, pp. 2594–2598.
[16] S. Ntalampiras, N. Fakotakis, Modeling the temporal evolution of acoustic parameters for speech emotion recognition, IEEE Trans. Affect. Comput. 3 (1) (2012) 116–125.
[17] L. Zão, D. Cavalcante, R. Coelho, Time-frequency feature and AMS-GMM mask for acoustic emotion classification, IEEE Signal Process. Lett. 21 (5) (2014) 620–624.
[18] J. Chang, X. Zhang, Q. Zhang, Y. Sun, Investigating duration effects of emotional speech stimuli in a tonal language by using event-related potentials, IEEE Access 6 (2018).
[19] Y. Kang, A. Li, Prosody conversion from neutral speech to emotional speech, IEEE Trans Audio Speech Lang. Process. 14 (4) (2006) 1145–1154.
[20] C. Wu, C. Hsia, C. Lee, M. Lin, Hierarchical prosody conversion using regression-based clustering for emotional speech synthesis, IEEE Trans. Audio Speech Lang. Process. 18 (6) (2010) 1394–1405.
[21] J. Jia, S. Zhang, F. Meng, Y. Wang, L. Cai, Emotional audio-visual speech synthesis based on PAD, IEEE Trans. Audio Speech Lang. Process. 19 (3) (2011) 570–582.
[22] N. Dey, A.S. Ashour, Direction of Arrival Estimation and Localization of Multi-Speech Sources, Springer International Publishing, 2018.
[23] N. Dey, A.S. Ashour, Applied examples and applications of localization and tracking problem of multiple speech sources, in: Direction of Arrival Estimation and Localization of Multi-Speech Sources, Springer, Cham, 2018, pp. 35–48.
[24] E. Navas, I. Hernaez, I. Luengo, An objective and subjective study of the role of semantics and prosodic features in building corpora for emotional TTS, IEEE Trans. Audio Speech Lang. Process. 14 (4) (2006) 1117–1127.
[25] S. Zhang, S. Zhang, T. Huang, W. Gao, Speech emotion recognition using deep convolutional neural network and discriminant temporal pyramid matching, IEEE Trans. Multimedia 20 (6) (2018) 1576–1590.
[26] F. Weninger, F. Eyben, B. Schuller, On-line continuous-time music mood regression with deep recurrent neural networks, in: IEEE International Conference on Acoustics, Speech and Signal Processing (ICASSP), Florence, 2014, 2014, pp. 5412–5416.

[27] A.R. Avila, J. Monteiro, D. O'Shaughneussy, T.H. Falk, Speech emotion recognition on mobile devices based on modulation spectral feature pooling and deep neural networks, in: IEEE International Symposium on Signal Processing and Information Technology (ISSPIT), Bilbao, 2017, 2017, pp. 360–365.

[28] N. Dey, A.S. Ashour, Challenges and future perspectives in speech-sources direction of arrival estimation and localization, in: Direction of Arrival Estimation and Localization of Multi-Speech Sources, Springer, Cham, 2018, pp. 49–52.

[29] N. Dey, A.S. Ashour, Sources localization and DOAE techniques of moving multiple sources, in: Direction of Arrival Estimation and Localization of Multi-Speech Sources, Springer, Cham, 2018, pp. 23–34.

[30] N. Dey, A.S. Ashour, Microphone array principles, in: Direction of Arrival Estimation and Localization of Multi-Speech Sources, Springer, Cham, 2018, pp. 5–22.

[31] Y. Zong, W. Zheng, T. Zhang, X. Huang, Cross-corpus speech emotion recognition based on domain-adaptive least-squares regression, IEEE Signal Process. Lett. 23 (5) (2016) 585–589.

[32] B. Sivakumar, et al., River flow forecasting: use of phase-space reconstruction and artificial neural networks approaches, J. Hydrol. 265 (2002) 225–245.

[33] S.P. Chandrasekaran, A nonlinear dynamic modelling for speech recognition using recurrence plot—a dynamic Bayesian approach, in: IEEE International Conference on Signal Processing and Communications, ICSPC 2007, 2007, pp. 516–519.

[34] A.M. Fraser, H.L. Swinney, Independent coordinates for strange attractors from mutual information, Phys. Rev. A 33 (1986) 1134.

[35] A. Dey, et al., A new kind of dynamical pattern towards distinction of pre-meditative and meditative states through HRV, Science 3 (2012).

[36] M.B. Kennel, H.D. Abarbanel, False neighbors and false strands: a reliable minimum embedding dimension algorithm, Phys. Rev. E 66 (2002) 026209.

[37] A. Dey, D.K. Bhattacha, D.N. Tibarewala, N. Dey, A.S. Ashour, D.-N. Le, E. Gospodinova, M. Gospodinov, Chinese-chi and kundalini yoga meditations effects on the autonomic nervous system: comparative study, Int. J. Interact. Multimedia Artif. Intell. 3 (7) (2016) 87–95. 9 p.

[38] M.B. Kennel, et al., Determining embedding dimension for phase-space reconstruction using a geometrical construction, Phys. Rev. A 45 (1992) 3403.

[39] J.P. Eckmann, et al., Recurrence plots of dynamical systems, EPL (Europhys. Lett.) 4 (1987) 973.

[40] N. Marwan, et al., Recurrence plots for the analysis of complex systems, Phys. Rep. 438 (2007) 237–329.

[41] S.K. Nayak, K. Pande, P.K. Patnaik, S. Nayak, S.J. Patel, A. Anis, A. Dey, K. Pal, Understanding the effect of cannabis abuse on the ANS and cardiac physiology of the Indian women paddy-field workers using RR interval and ECG signal analyses. in: IEEE International Conference APSIPA ASC 2017, Aloft Kuala Lumpur Sentral Sentral, Kuala Lumpur, 2017. https://doi.org/10.1109/APSIPA.2017.8282047.

[42] A. Dey, S.K. Palit, S. Mukherjee, D.K. Bhattacharya, D.N. Tibarewala, A new technique for the classification of pre-meditative and meditative states, in: IEEE International Conference, "ICCIA-2011", 2011. Print ISBN 978-1-4577-1915-8.

[43] M. Das, T. Jana, P. Dutta, R. Banerjee, A. Dey, D.K. Bhattacharya, M.R. Kanjilal, Study the effect of music on HRV signal using 3D poincare plot in spherical co-ordinates—a signal processing approach, in: IEEE International Conference on Communication and Signal Processing, April 2–4, 2015, India, 2015. ISBN 978-1-4799-8080-2.

[44] H. Sabelli, A. Lawandow, Homeobios: the pattern of heartbeats in newborns, adults, and elderly patients, in: Nonlinear Dynamics, Psychology, and Life Sciences, vol. 14, 2010, p. 381.

Chapter 9

Intelligent Speech Processing in the Time-Frequency Domain

Biswajit Karan, Kartik Mahto and Sitanshu Sekhar Sahu
Department of Electronics and Communication Engineering, Birla Institute of Technology, Mesra, Ranchi, India

9.1 Wavelet Packet Decomposition

Wavelets packet decomposition is an effective technique for analyzing a signal that is nonstationary in nature, especially for speech signal. It is computationally more efficient and performs well in comparison to the short-time signal processing technique like Fourier transform methods. In wavelet packet transformation (WPT) analysis, the variable window size is applied to captured high and low frequency band information. It represents both high pass and low pass results as a generalization of wavelet decomposition. WPT decomposes the signal into sub-bands, which provides good time and frequency resolution. Wavelet packets are waveforms indexed by three parameters: position, scale, and frequency.

A wavelet packet tree is associated with a level j, which splits into two bands by decomposing both low- and high-frequency components called approximation and detail coefficients, respectively. The result of this decomposition is a balanced tree structure.

The computation of wavelet packet generation starts with scaling filter $h(n)$ as the low pass filter and wavelet filter as the high pass filter each of length $2N$. We start the decomposition with the two filters of length $2N$, where $h(n)$ and $g(n)$ correspond to filter

$$W_{2n}(x) = \sqrt{2}\sum_{k=0}^{2N-1} h(k)W_n(2x-k) \tag{9.1}$$

$$W_{2n+1}(x) = \sqrt{2}\sum_{k=0}^{2N-1} g(k)W_n(2x-k) \tag{9.2}$$

$W_0(x) = \varphi(x)$ is the scaling function and $W_1(x) = \psi(x)$ is the wavelet function.

Intelligent Speech Signal Processing. https://doi.org/10.1016/B978-0-12-818130-0.00009-X

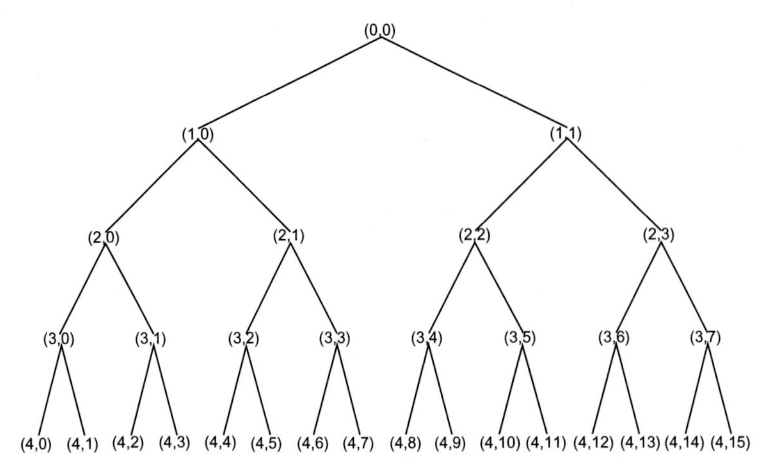

FIG. 9.1 Wavelet decomposition tree of depth 4.

Starting from the initial functions ($W_n(x)$, $n \in N$), the three-indexed wavelet functions are given by

$$W_{j,n,k}(x) = 2^{-j/2} W_n\left(2^{-j/2} x - k\right) \tag{9.3}$$

where $n \in N$ and $(j, k) \in Z^2$.

As in the wavelet decomposition, k is the time-localized parameter and j as a scale parameter. The set of functions ($W_{j,\ n} = W_{j,\ n,\ k}(\mathrm{x})$, $\mathrm{k} \in z$) represents the (j, n) wavelet packet. j and n, are the positive integers. Wavelet packets are arranged in trees (Fig. 9.1).

The same speech signal can be represented in terms of wavelet packet of depth $j = 4$. So after decomposition, we get 16 wavelet packets (Fig. 9.2). This wavelet packet can be utilized by further processing to get an efficient characterization of the speech signal. We can take advantage of the wavelet packet by taking a different feature like the statistical feature (mean, variance, kurtosis, and skewness), energy, and entropy of wavelet packet. These features can be utilized in pathological speech classification, speaker verification, automatic speech recognition (ASR), emotion recognition, and gender classification.

In wavelet analysis, a signal is fragmented into an approximate, and detailed coefficients are further filtered and downsampled. The approximation is then itself split into a second-level approximation and detail, and the process is repeated for more precise coefficients, and so on. Mallat [1] provides a thorough discussion of wavelet analysis (Fig. 9.3).

9.1.1 Spectral Analysis

The tree structure of wavelet analysis gives the accurate estimation of the frequency band of speech perception better than other methods. Abrupt variation of the speech signal in both time and frequency domain can be tracked using

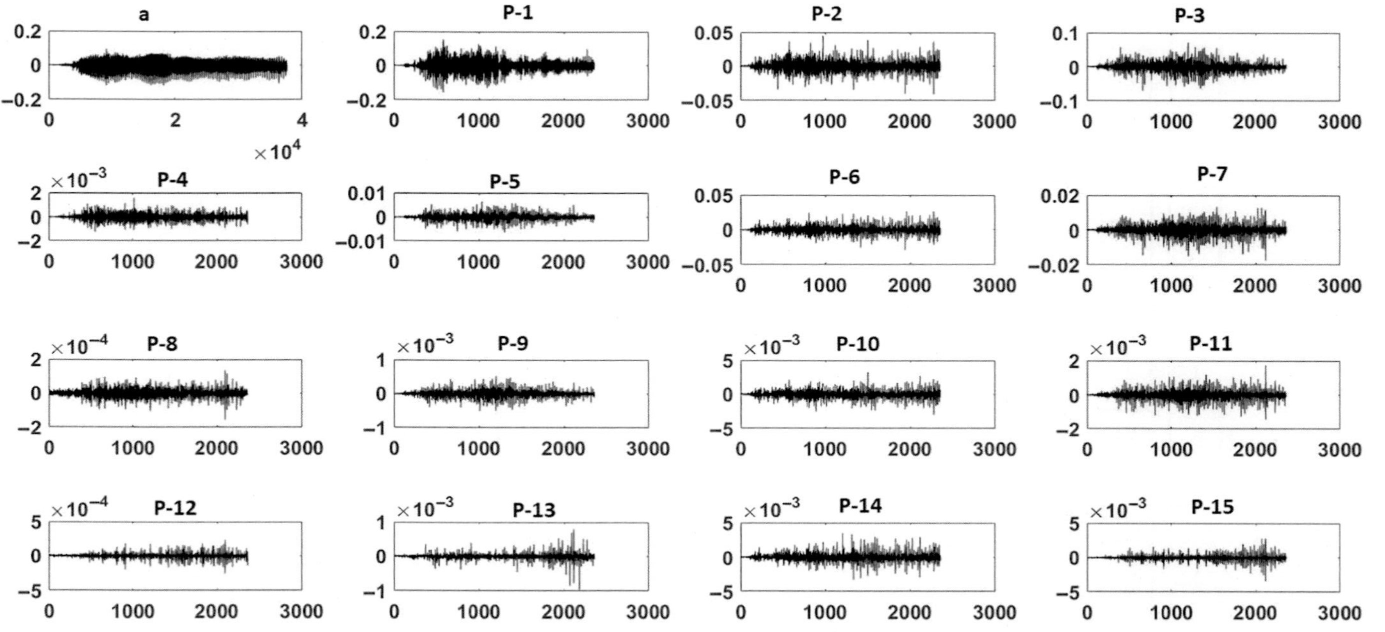

FIG. 9.2 (A) The original speech signal and (P-1) to (P-15) represents a decomposed wavelet packet.

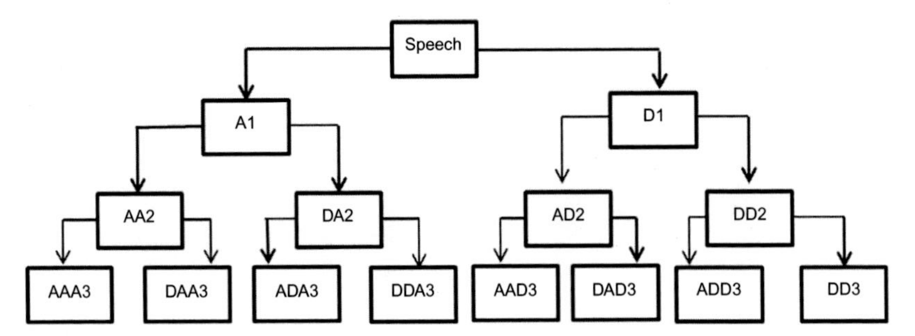

FIG. 9.3 Structure of three-level wavelet packet trees.

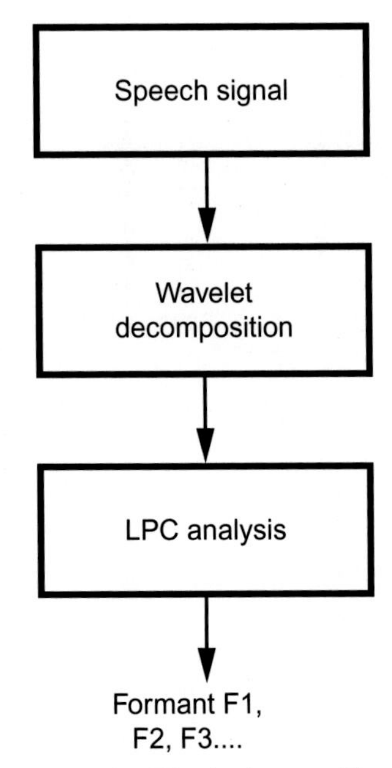

FIG. 9.4 Formant estimation steps using Wavelet decomposition of the speech signal.

WPT. This wavelet packet can be utilized successfully for pitch detection, voiced and unvoiced classification of the speech signal. Formant estimation can also be done using wavelet analysis. Many algorithms have been developed for formant estimation [2–4]. The formant or vocal tract information can be found using a combined approach of wavelet and linear predictive coding (LPC) analysis. First, taking a speech signal and estimating the wavelets, then taking the linear predictive coefficient spectrum provides the formant frequency of speech signal. It gives an accurate estimation of pitch as compared with the traditional algorithm (Fig. 9.4).

9.1.2 Pitch Detection

Pitch detection is an important task in speech processing. It is defined as a fundamental frequency of speech signal generated from vocal fold vibration. Researchers proposed many traditional pitch detection algorithms such as spectral and time domain.

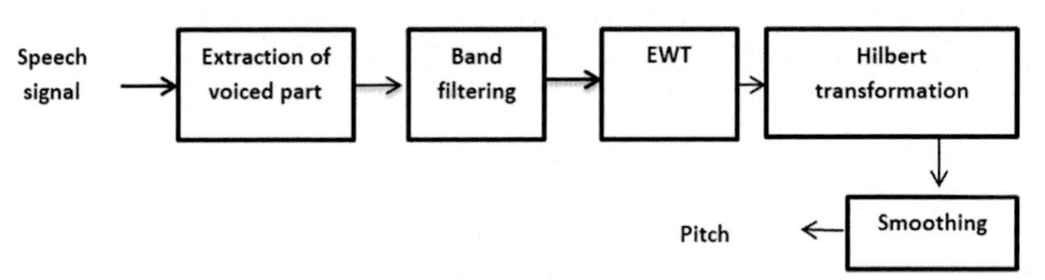

FIG. 9.5 Pitch detection using empirical wavelet analysis.

The spectral-based pitch detection algorithm is based on autocorrelation and the cepstral method. The time-based pitch detector is measured between two successive glottal closure instants (GCIs). Research focused on several pitch detection algorithms based on wavelet analysis in [5–8]. Recently, empirical wavelet analysis was proposed for accurate estimation of pitch [6]. The following block diagram shows the estimation of the pitch using empirical wavelet transformation (Fig. 9.5).

Researchers used common applications for WPT, such as voice signal processing, image processing, and pathological speech processing.

9.2 Empirical Mode Decomposition

Empirical mode decomposition (EMD) is an adaptive decomposition technique for the nonstationary type of signal. N. Huang introduced the technique in 1998. It decomposed the speech signal into AM-FM (amplitude modulation and frequency modulation) components called modes or intrinsic mode functions (IMFs). Sharma et al. [9] have been effectively utilized EMD Algorithm to characterized the speaker specific information. The EMD algorithm is summarized in the following steps:

(a) Identify local extrema between two successive zero crossings of signal $x(t)$.
(b) Join all maxima by cubic spline line considered as upper envelope [E1].
(c) Repeat the above for a lower envelope [E2].
(d) Calculate the mean of the upper and lower envelope [$m = (\mathrm{E1} + \mathrm{E2})/2$].
(e) The first IMFs are given by $h_1(t) = x(t) - m$.

The IMFs should satisfy the following criteria: The number of extrema and zero crossing should be the same and its envelope should be symmetric with respect to zero.

For a signal, the decomposition into IMFs may be represented as

$$s(n) = r_k(n) + \sum_{i=1}^{k} c_i(n) \tag{9.4}$$

where $r_k(n)$ is the residue and $c_i(n)$ and IMF of ith mode.

It is a data-adaptive technique that decomposed and effectively characterized the speech signal. The speaker-specific information was well distributed among modes or IMFs.

Key Points of EMD

1. It effectively decomposed the nonstationary signal into IMFs or modes. These modes are time domain components that give the same signal after the addition of all IMFs. It is a complete and reliable tool for signal decomposition.
2. It gives a meaningful component that carries both vocal tract (formant) and vocal fold information (pitch).
3. It overcomes the disadvantages of short-time processing.
4. Because of dyadic filter-bank nature of EMD, it effectively decompose the signal into modes or intrinsic mode function.

9.2.1 EMD of Synthetic and Speech Signal

For example, take a synthetic signal of fundamental frequency 300 Hz and formant frequencies 900, 1800, 2400, 3000 Hz. After decomposition, it is converted into four numbers of IMFs (Figs. 9.6 and 9.7).

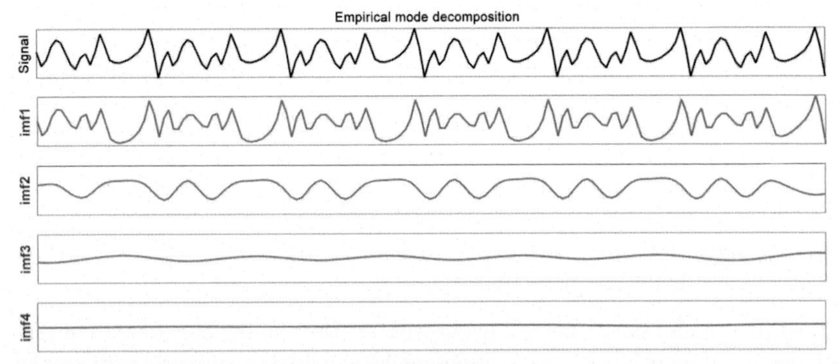

FIG. 9.6 Synthetic signal and four decomposed IMFs.

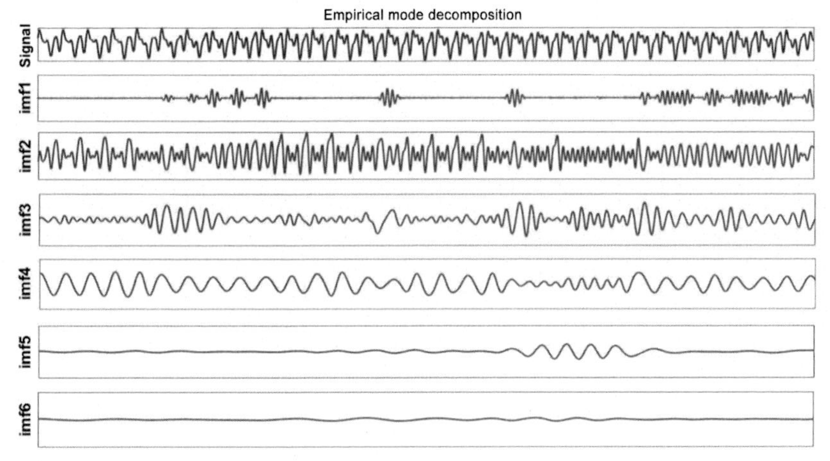

FIG. 9.7 Speech signal and decomposed IMFs.

9.2.2 Spectrum of IMFs

The speaker-specific information is distributed among decomposed IMFs. The vocal tract and vocal fold information can be obtained from successive IMFs. A different analysis shows that the first four IMFs consist of vocal tract information of speaker [9, 10].

Fig. 9.8 shows that IMFs can be used to find the vocal tract information of speaker, and the spectrum of IMFs gives the formant frequency.

9.2.3 Estimation of Pitch

Researchers have been proposed different pitch tracking algorithms [7, 8]. But EMD-based GCI is more efficient than traditional algorithm. G. Schlotthauer et al. [11] proposed an EMD-based algorithm for pitch estimation. After IMFs are extracted from EMD, the primary objective is to select appropriate IMFs that hold information about the pitch. The author's inspection showed that pitch tracking information is found in fifth, sixth and seventh IMFs. He proposed entropy-based selection to pinpoint the appropriate IMFs. It was found that fifth IMFs carried the most information about the pitch. That is confirmed by applying a threshold-based approach for selecting appropriate IMFs. Further research is needed for the accurate study of IMFs. Another approach has been proposed by Sharma et al. [10] based on EMD. He proposed a methodology for estimation of GCI. A combination of IMFs were taken for detection of GCI. The partial summation of IMFs provides a sound estimation of GCI (Fig. 9.9). The partial summation gives the sinusoidal like waveform, which enables the detection of the GCI. Two consecutive GCI gives the pitch information of signal [10].

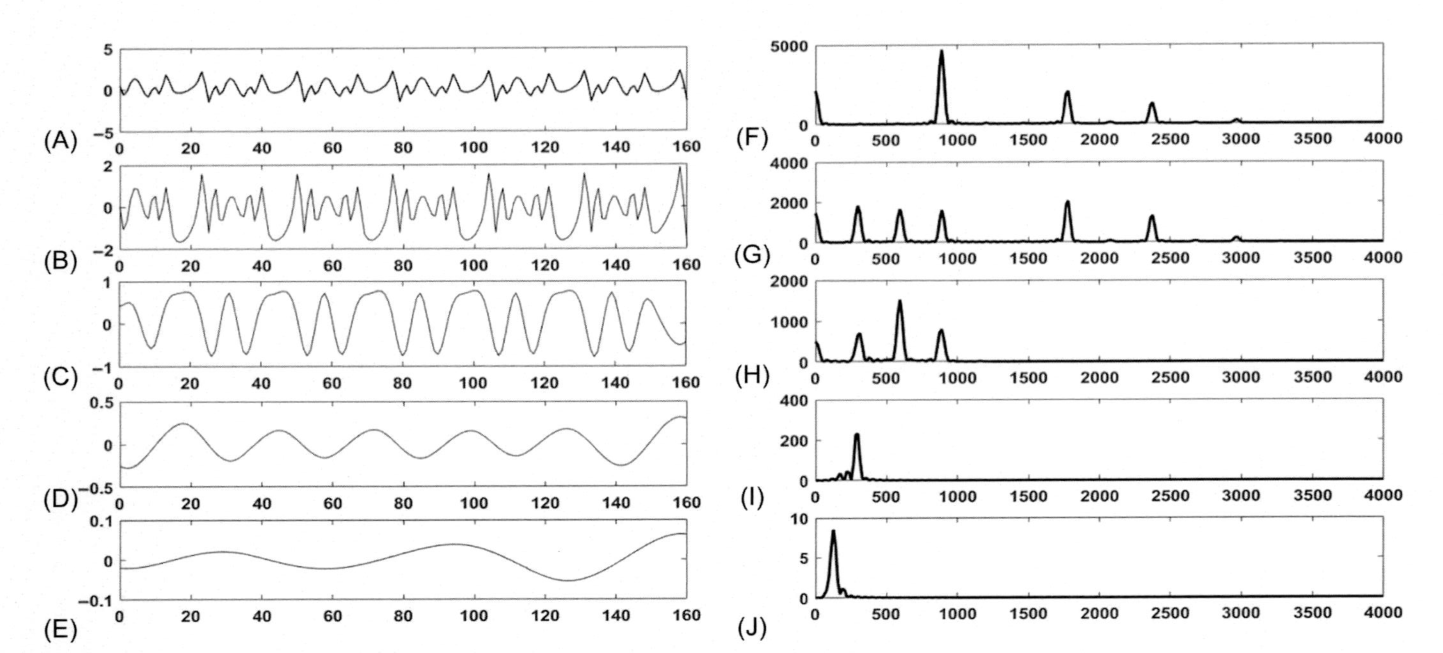

FIG. 9.8 Formant distribution in the IMFs (A) Synthetic speech voice signal (B)–(E) decomposed IMFs. (F) The spectrum of synthetic signal peaks at frequency 900, 1800, 2400 and 3000 Hz. (G)–(J) Spectrum of IMFs.

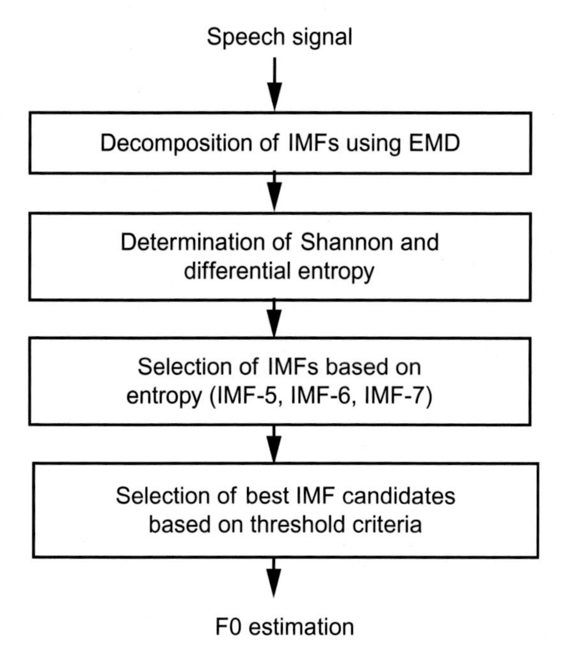

FIG. 9.9 Steps for estimation of the pitch from IMFs.

9.3 Variational Mode Decomposition

Variational mode decomposition (VMD) breaks down a real-valued signal into a finite number of subsignals called modes, μ_k and corresponding center frequency, w_k [12]. Each mode or subsignal occurred around center frequency [13]. Each mode is extracted using the following steps:

- To find a one-sided spectrum of each mode using the Hilbert transform.
- Shift each mode spectrum using modulation property by multiplying $e^{-jw_k n}$.
- Find the mode bandwidth by Gaussian smoothing.

The VMD is implemented by solving the following constrained optimization problem:

$$\min_{\{u_k\},\{w_k\}} \left\{ \sum_k \left\| \partial_t \left[\left(\delta(t) + \frac{j}{\pi t} \right) \times \mu_k(t) \right] e^{-jw_k t} \right\|_2^2 \right\} \tag{9.5}$$

$$\text{Such that} \sum_k u_k = f$$

where u_k is the kth decomposed mode; w_k, the center frequency of kth mode signal; $f(t)$, input signal; and $\left[\left(\delta(t) + \frac{j}{\pi t} \right) \times \mu_k(t) \right]$ is the Hilbert transform of the exponential term shift the frequency spectrum of each mode into the baseband. Using the augmented Lagrangian method, L, above minimization problem can be solved as follows:

$$L(\{u_k\},\{w_k\},\lambda)=\alpha\sum_k\left\|\partial_t\left[\left(\delta(t)+\frac{j}{\pi t}\right)\times\mu_k(t)\right]e^{-jw_kt}\right\|_2^2 + \left\|f(t)-\sum_k u_k(t)\right\|_2^2+\left\langle\lambda(t),f(t)-\sum_k u_k(t)\right\rangle \tag{9.6}$$

The complete algorithm for VMD known as ADMM (alternating direction method of multiplier) [12] is summarized as follows.

ADMM Optimization Algorithm for VMD

Initialize $\{u_k^1\}$, $\{w_k^1\}$, λ^1, $n\leftarrow 0$

Repeat $n\leftarrow n+1$

for $k=1{:}K$ do

Update: u_k

$$u_k^{n+1}\leftarrow \arg_k\min L\left(\{u_{i>k}^{n+1}\},\{u_{i\geq k}^{n}\},\{w_i^n\},\lambda^n\right) \tag{9.7}$$

end for

for $k=1{:}K$ do

Update: w_k

$$w_k^{n+1}\leftarrow \arg_k\min L\left(\{u_i^{n+1}\},\{w_{i<k}^{n}\},\{w_{i\geq k}^{n}\},\lambda^n\right) \tag{9.8}$$

end for

Dual ascent:

$$\lambda^{n+1}\leftarrow\lambda^n+\tau\left(f-\sum_k u_k^{n+1}\right) \tag{9.9}$$

Until convergence:

$$\sum_k \left\|u_k^{n+1}-u_k^n\right\|_2^2 \Big/ \left\|u_k^n\right\| 2^2<\varepsilon \tag{9.10}$$

The number of the parameter should be initialized before decomposition. For decomposition, a number of modes $k=5$, data fidelity constrained, $\alpha=120$ and tolerance convergence criteria, tol $=10^{-7}$ is taken for cold speech decomposition [14]. The parameter can be changed for another application area.

Fig. 9.10 shows the VMD of the synthetic voice signal of fundamental frequency 300 Hz and formant frequency 900, 1800, 2400, and 3000 Hz and corresponding spectrum. Corresponding subsignal or modes capture the vocal fold and vocal tract information very effectively. VMD is more advantageous and efficient than EMD. EMD is a recursive method, whereas VMD is nonrecursive in nature. EMD has several disadvantages like a lack of mathematical expression and sensitive to noise.

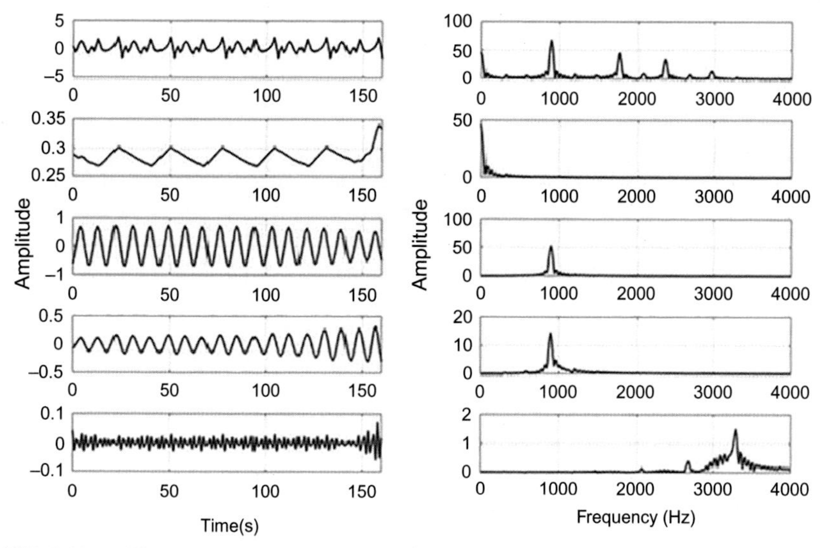

FIG. 9.10 Different mode signals of the synthetic voice signal and its spectrum.

9.3.1 Voiced and Unvoiced Detection of Speech

A. Upadhya et al. [13] proposed an instantaneous detection of voiced/nonvoiced regions of a speech signal and EGG (electroglottography) using VMD-based technique. The VMD is found in an iterative way with some input parameters. In the first steps, the VMD split the speech signal into two modes; then again VMD is applied on one of these two selected components based on some convergence criteria. A positive separation of the fundamental frequency (F_0) component from the speech signal was found.

9.3.2 Estimation of Pitch Period

Many algorithms have been proposed for pitch estimation using a traditional technique like autocorrelation and cepstral method. Using VMD, a pitch period was found more accurately. It is proposed by Upadhyay et al. [15] (Fig. 9.11).

9.4 Synchrosqueezing Wavelet Transform: EMD Like a Tool

Synchrosqueezing wavelet transform (SST) is a nonlinear, time-frequency method based on continuous wavelet transformation (CWT). It distributes the energy of signal into frequency. It mitigates the effect of the spreading of the mother wavelet. SST reassigned energy into a frequency direction, which conserves the time resolution of the signal. So, it better reconstructs the signal. A signal with a strong noise, SST gives clean time-frequency representation. It reduces the effect of mode mixing [16]. The SST algorithm is used in many

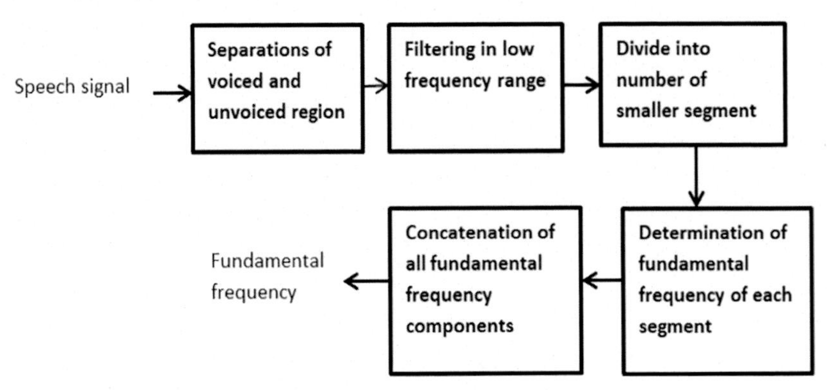

FIG. 9.11 Estimation of fundamental frequency using VMD.

applications like climate analysis, mechanical fault diagnosis, and signal demising. The real world signal can be represented as the AM-FM component and given by the following equation:

$$x(t) = \sum_{p=1}^{N} A_p(t) \cos\left[2\pi\varphi_p(t)\right] \tag{9.11}$$

where $A_p(t)$ is the slowly varying amplitude signal and $\varphi_p(t)$ is the instantaneous phase of the signal.

Algorithm

Step 1: From input signal, first find the CWT.
Step 2: Find the instantaneous frequency from CWT output ($W_f(s, u)$)

$$w_f(s,u) = \frac{\partial t W_f(s,u)}{2\pi i W_f(s,u)} \tag{9.12}$$

The phase transformation is proportional to first derivative CWT and scale s is defined as

$$s = \frac{f_x}{f} \tag{9.13}$$

where f_x is the peak frequency and f is the frequency.

To calculate the instantaneous frequency, consider a simple sine wave, $e^{j2\pi f_0}$

✓ Calculate CWT: $W_f(e^{j2\pi f_0 u}) = e^{j2\pi f_0 u}$
✓ Take partial derivative:

$$\frac{\partial}{\partial u} W_f\left(e^{j2\pi f_0 t}\right) = i2\pi f_0 \hat{\chi}\left(f_\chi\right) e^{j2\pi f_0 u} \tag{9.14}$$

where $\hat{\chi}\left(f_\chi\right)$ is the Fourier transform at sf_0.

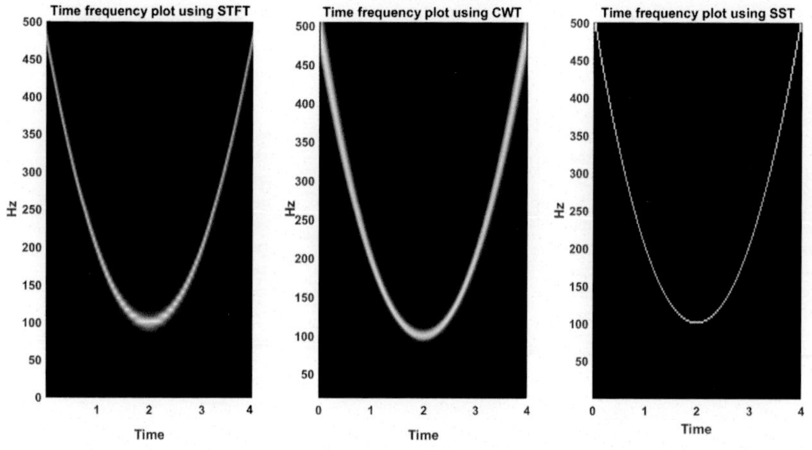

FIG. 9.12 Comparison of time-frequency plot of STFT, CWT, and SST of the quad chirp signal.

✓ To get the instantaneous frequency, divide the partial derivative by $i2\pi$ and wavelet transformation.

Step 3: Squeeze the CWT over the region where there is a constant phase transformation value. The instantaneous frequency value is converted into single value. The result is sharper compared to a short-time Fourier transform (STFT) and CWT [17] (Fig. 9.12).

9.4.1 Reconstructions of Speech Signal Using Synchrosqueezed Transform

We can reconstruct a signal from the synchrosqueezed transform. This is an advantage synchrosqueezing transform has over other time-frequency reassignment techniques. The transform does not provide a perfect inversion, but the reconstructed signal is stable and the results are typically good [16]. The ability to reconstruct from the synchrosqueezed transform enables us to extract signal components from localized regions of the time-frequency plane. Fig. 9.13 is showing the extraction of the original signal from modes. The reconstruction of an original signal from individual modes is not exact, but the ability provides good time-frequency localization as compared to CWT and STFT. It is the best method for analyzing the signal in a time-frequency domain.

9.4.2 Advantage of Synchrosqueezed Wavelet Transforms

✓ Sharpen time-frequency analysis: Synchrosqueezing can compensate for the spreading in time and frequency caused by linear transforms like the STFT and CWT. In CWT, the wavelet acts like a measuring device for the input signal.

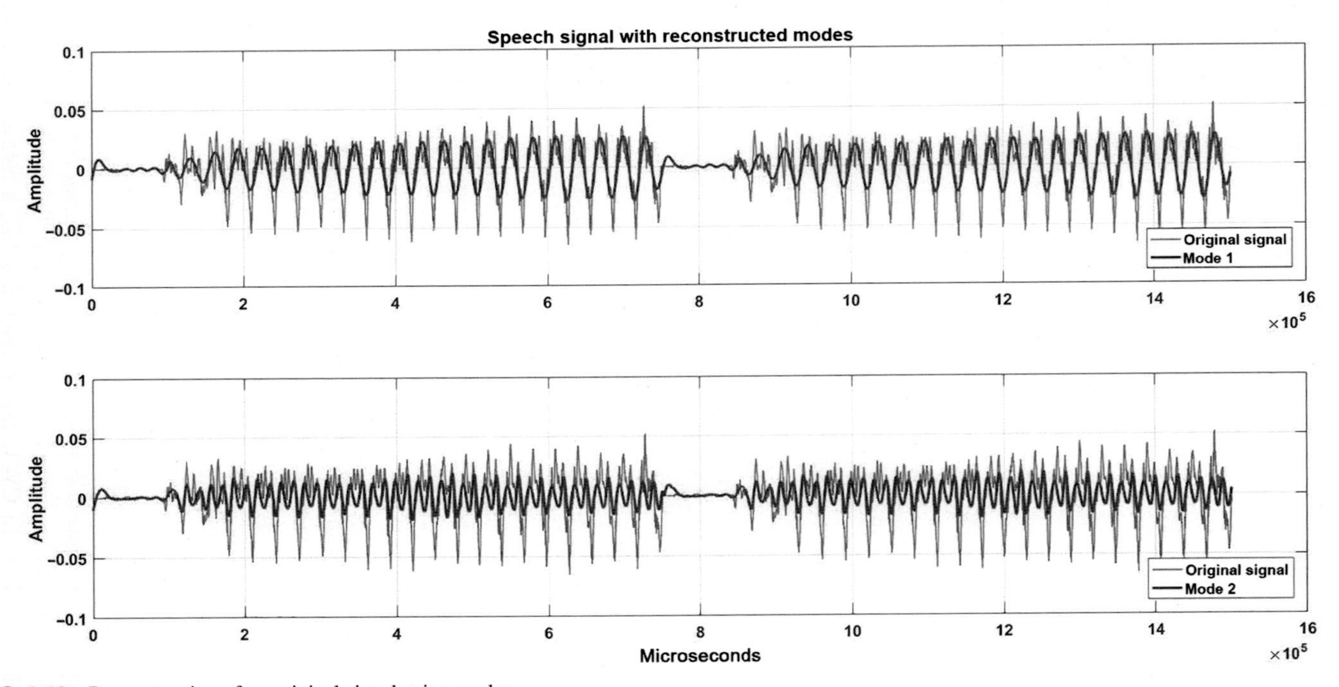

FIG. 9.13 Reconstruction of an original signal using modes.

- ✓ The transform does not provide a perfect inversion, but the reconstruction is stable and the results are typically quite good.
- ✓ It reduced the effect of mode mixing.

9.5 Applications of the Decomposition Technique

In this chapter, we have discussed four nonlinear decomposition techniques.

- ✓ Wavelet packet analysis
- ✓ Empirical mode decomposition
- ✓ Variational mode decomposition
- ✓ Synchrosqueezed wavelet transform

9.5.1 Feature Extraction

For the above technique, several features are extracted from decomposed time domain subsignals. The following features are extracted from modes.

1. *Statistical parameter:* Speech has good pitch and amplitude variation from person to person. These variations are well defined by using mean, variance, skewness, and kurtosis of a different mode.

2. *Energy:* The basic feature energy of each mode is captured for each speech signal. It is given by

$$E_{u_i} = \frac{1}{N}\sum_{i=1}^{N}(u_i)^2 \tag{9.15}$$

where u_i is the time domain signal of the ith mode and N is the number of samples of each mode.

   Energy is normalized by dividing it with N to remove the dependency on the sample length.

3. *Energy entropy:* The short-term entropy of energy can be interpreted as a measure of abrupt changes in the energy level of a speech signal.

It is defined as

$$H(i) = -\sum_{j=1}^{k} e_j \cdot \log_2(e_j) \tag{9.16}$$

$$j = 1, 2 \ldots k \text{ and } i = \text{No.of modes}$$

where $e_j = \frac{E_{subfram\,e_j}}{E_{shortfram\,e_i}}$ and $E_{shortframe_i} = \sum_{k=1}^{K} E_{subframe_k}$.

These features of the extracted mode are used in many application areas like speaker verification and pathological speech processing [9]. Sharma et al. reported that EMD-based method has efficiently increased the performance of the speaker verification compared to a traditional feature like mel frequency cepstral coefficient (MFCC). The VMD-based feature has proven better for the classification of cold speech and pathological speech [14].

9.5.2 Clinical Diagnosis and Pathological Speech Processing

This nonlinear decomposition has great importance in the field of speech disorder like hypernasality, breathiness, and dysphonia [18–22]. The first speech signal is recorded from the patient. It is preprocessed then decomposed in modes using the nonlinear technique. These modes are efficiently characterized as speaker-specific information (Fig. 9.14).

Then, different types of features are extracted such as energy, entropy, and statistical features (mean, median, kurtosis, skewness) [14, 23, 24]. Now using different classifier we able to distinguish between pathological and a healthy voice [23]. In case of vocal fold related problems, common characteristics such as reduced pitch decreased in the intensity of pitch, and irregular energy distribution in time-frequency representation [25].

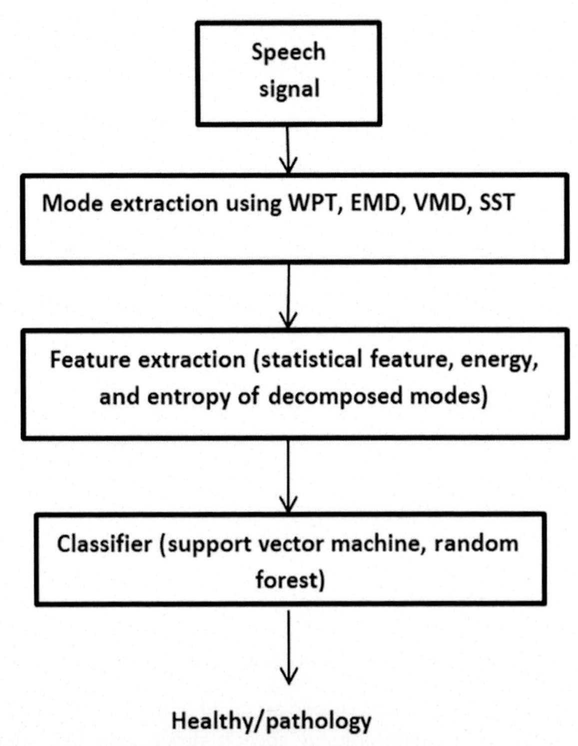

FIG. 9.14 Flow diagram of detection of pathological speech.

9.5.3 Automatic Speech Recognition

Many researchers have proposed ASR system using a traditional feature like MFCC, LPC, linear predictive cepstral coefficient (LPCC), and PLP (perceptual linear prediction) [26–28]. It is challenging to develop an automatic speech recognizer in presence of noise. These signal decomposition techniques are a good tool for denosing the speech signal [29–31]. To develop a voiced-enabled device, the ASR is key. For the development of ASR, we need a large database [31]. Then, we extracted the feature using a different decomposition technique [32, 33]. Using the hidden Markov model-based recognizer, we get the output.

Fig. 9.15 shows the basic speech recognition system. Currently, the primary focus worldwide is on the design of speech-based access of the automated system. So, we need a robust feature that recognized the speech signal in every situation.

9.5.4 Other Application Area

These nonlinear techniques are useful in speech processing, localization, and tracking of acoustic sources [34–36]. Joint-time frequency analysis, is useful in the tracking of an audio-based camera steering system. These nonlinear

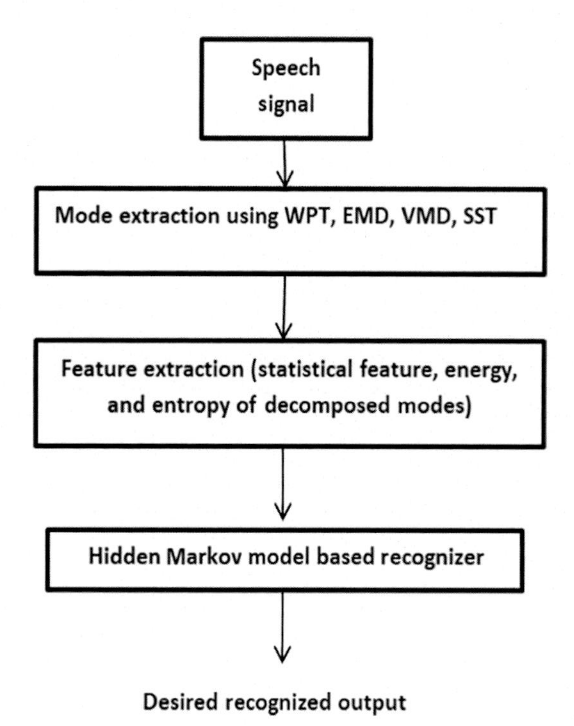

FIG. 9.15 Flow diagram of the ASR system.

techniques are useful in many signals processing areas compared to the automatic direction of arrival estimation (DOAE) technique [36–38].

9.6 Conclusion

In this chapter, we have described the different nonlinear decomposition techniques. We have showed that each decomposition technique has its advantages and disadvantages. The wavelet decomposition technique better characterized the speaker-specific information than continuous and discrete wavelet transformation because of its variable window size. It captured both high- and low-frequency results as a generalization of wavelet decomposition. But it does not give both time and frequency resolution at the same time. It does not capture transient activity at low frequency or rhythmic activity at high frequency. Thus this problem is overcome by synchrosqueezed wavelet transform. It gives good time-frequency localization. The EMD is a nonlinear technique, which is able to capture information for speaker verification and pathological classification tasks. But because of the lack of mathematical proof and the mode mixing problem, it fails in some application area. The VMD is another technique introduced in this chapter for effective decomposition of the signal. This technique proves better in some application area like cold speech classification, voiced and unvoiced detection. It has enough mathematical justification. These decomposition techniques have a great role in time-frequency analysis of speech signal.

References

[1] S. Mallet, A Wavelet Tour of Signal Processing, Academic Press, San Diego, CA, 2008, p. 89.
[2] I. Jemaa, K. Ouni, Y. Laprie, Automatic formant tracking method by Fourier and multiresolution analysis, IFAC Proc. 43 (8) (2010) 347–352.
[3] M. Bensaid, J. Schoentgen, S. Ciocea, Estimation of formant frequencies by means of a wavelet transform of the speech spectrum, in: Proceeding of the ProRISC Workshop on Circuits, Systems and Signal Processing, 1997, pp. 42–46.
[4] D.Y. Loni, S. Subbaraman, Formant estimation of speech and singing voice by combining wavelet with LPC and Cepstrum techniques, in: 2014 9th International Conference on Industrial and Information Systems (ICIIS), Gwalior, 2014, pp. 1–7.
[5] J.F. Wang, S.H. Chen, J.S. Shyuu, Wavelet transforms for speech signal processing, J. Chin. Inst. Eng. 22 (5) (1999) 549–560.
[6] T.S. Kumar, M.A. Hussain, V. Kanhangad, Classification of voiced and non-voiced speech signals using empirical wavelet transform and multi-level local patterns, in: 2015 IEEE International Conference on Digital Signal Processing (DSP), Singapore, 2015, pp. 163–167.
[7] D. Charalampidis, V.B. Kura, Novel wavelet-based pitch estimation and segmentation of non-stationary speech, in: 8th Int. Conf. Inf. Fusion, vol. 2, 2005, pp. 1383–1387.
[8] M. Obaidat, C. Lee, B. Sadoun, D. Neslon, Estimation of pitch period of speech signal using a new dyadic wavelet transform, J. Inf. Sci. 119 (1999) 21–39.
[9] R. Sharma, S. Prasanna, R.K. Bhukya, R.K. Das, Analysis of the intrinsic mode functions for speaker information, Speech Comm. 91 (2017) 1–16.

[10] R. Sharma, S.R.M. Prasanna, H.L. Rufiner, G. Schlotthauer, Detection of the glottal closure instants using empirical mode decomposition, Circuits Systems Signal Process. 37 (8) (2017) 3412–3440.
[11] G. Schlotthauer, M.E. Torres, H.L. Rufiner, Voice fundamental frequency extraction algorithm based on ensemble empirical mode decomposition and entropies, in: World Congress on Medical Physics and Biomedical Engineering, September 7–12, 2009, Munich, Germany, Springer, Berlin, Heidelberg, 2009.
[12] K. Dragomiretskiy, D. Zosso, Variational mode decomposition, IEEE Trans. Signal Process. 62 (3) (2014) 531–544.
[13] A. Upadhyay, R.B. Pachori, Instantaneous voiced/non-voiced detection in speech signals based on variational mode decomposition, J. Franklin Inst. 352 (7) (2015) 2679–2707.
[14] S. Deb, S. Dandapat, J. Krajewski, Analysis and classification of cold speech using variational mode decomposition, IEEE Trans. Affect. Comput. (2017).
[15] A. Upadhyay, M. Sharma, R.B. Pachori, Determination of instantaneous fundamental frequency of speech signals using variational mode decomposition, Comput. Electr. Eng. 62 (2017) 630–647.
[16] I. Daubechies, J. Lu, H.T. Wu, Synchrosqueezed wavelet transforms: an empirical mode decomposition-like tool, Appl. Comput. Harmon. Anal. 30 (2) (2011) 243–261.
[17] G. Thakur, E. Brevdo, N.S. Fučkar, H.T. Wu, The synchrosqueezing algorithm for time-varying spectral analysis: robustness properties and new paleo climate applications, Signal Process. 93 (2013) 1079–1094.
[18] A. Tsanas, et al., Novel speech signal processing algorithms for high-accuracy classification of Parkinson's disease, IEEE Trans. Biomed. Eng. 59 (5) (2012) 1264–1271.
[19] B.E. Sakar, et al., Collection and analysis of a Parkinson speech dataset with multiple types of sound recordings, IEEE J. Biomed Health Inform. 17 (4) (2013) 828–834.
[20] E.A. Belalcazar-Bolanos, et al., Automatic detection of Parkinson's disease using noise measures of speech, in: XVIII Symposium of IEEE Image, Signal Processing, and Artificial Vision (STSIVA), 2013, 2013.
[21] O. Arroyave, R. Juan, et al., Spectral and cepstral analyses for Parkinson's disease detection in Spanish vowels and words, Expert. Syst. 32 (6) (2015) 688–697.
[22] J.R. Orozco-Arroyave, et al., Automatic detection of Parkinson's disease in running speech spoken in three different languages, J. Acoust. Soc. Am. 139 (1) (2016) 481–500.
[23] R. Amami, A. Smiti, An incremental method combining density clustering and support vector machines for voice pathology detection, J. Comput. Electr. Eng. 57 (2017) 257–265.
[24] V. Parsa, D.G. Jamieson, Acoustic discrimination of pathological voice: sustained vowels versus continuous speech, J. Speech Lang. Hear. Res. 44 (2) (2001) 327–339.
[25] M.E. Torres, et al., Empirical mode decomposition. Spectral properties in normal and pathological voices, in: 4th European Conference of the International Federation for Medical and Biological Engineering, Springer, Berlin, Heidelberg, 2009.
[26] Z. Tufekci, J.N. Gowdy, S. Gurbuz, E. Patterson, Applied mel-frequency discrete wavelet coefficients and parallel model compensation for noise-robust speech recognition, Speech Comm. 48 (10) (2006) 1294–1307.
[27] O. Farooq, S. Datta, Wavelet-based denoising for robust feature extraction for speech recognition, Electron. Lett. 39 (1) (2003) 163–165.
[28] M. Gupta, A. Gilbert, Robust speech recognition using wavelet coefficient features, automatic speech recognition and understanding, in: IEEE Automatic Speech Recognition and Understanding Workshop 2001 (ASRU'01), Madonna di Campiglio, Italy, 2001, pp. 445–448.

[29] P.K. Sahu, A. Biswas, A. Bhowmick, M. Chandra, Auditory ERB like admissible wavelet packet features for TIMIT phoneme recognition, Eng. Sci. Technol. Int. J. 17 (3) (2014) 145–151.

[30] P.K. Astik Biswas, A.B. Sahu, M. Chandra, Feature extraction technique using ERB like wavelet sub-band periodic and aperiodic decomposition for TIMIT phoneme recognition, Int. J. Speech Technol. 17 (4) (2014) 389–399.

[31] G. Choueiter, J. Glass, An implementation of rational wavelets and filter design for phonetic classification, IEEE Trans. Audio Speech Lang. Process. 15 (3) (2007) 939–948.

[32] B. Rehmam, Z. Halim, G. Abbas, T. Muhammad, Artificial neural network- based speech recognition using DWT analysis applied on isolated words from oriental language, Malays. J. Comput. Sci. 28 (3) (2015) 242–262.

[33] E. Pavez, J.F. Silva, Analysis and design of wavelet-packet cepstral coefficients for automatic speech recognition, Speech Comm. 54 (6) (2012) 814–835.

[34] N. Dey, A.S. Ashour, W.S. Mohamed, N.G. Nguyen, Acoustic wave technology, in: Acoustic Sensors for Biomedical Applications, Springer, Cham, 2019, pp. 21–31.

[35] N. Dey, A.S. Ashour, W.S. Mohamed, N.G. Nguyen, Acoustic sensors in biomedical applications, in: Acoustic Sensors for Biomedical Applications, Springer, Cham, 2019, pp. 43–47.

[36] N. Dey, A.S. Ashour, Applied examples and applications of localization and tracking problem of multiple speech sources, in: Direction of Arrival Estimation and Localization of Multi-Speech Sources, Springer, Cham, 2018, pp. 35–48.

[37] N. Dey, A.S. Ashour, Direction of Arrival Estimation and Localization of Multi-Speech Sources, Springer, 2018.

[38] N. Dey, A.S. Ashour, Sources localization and DOAE techniques of moving multiple sources, in: Direction of Arrival Estimation and Localization of Multi-Speech Sources, Springer, Cham, 2018, pp. 23–34.

Further Reading

M. Hesham, A predefined wavelet packet for speech quality assessment, J. Eng. Appl. Sci. 53 (5) (2006) 637–652.

A. Karmakar, A. Kumar, R.K. Patney, A multi-resolution model of auditory excitation pattern and its application to objective evaluation of perceived speech quality, IEEE Trans. Audio Speech Lang. Process. 14 (6) (2006) 1912–1923.

W. Dobson, J. Yang, K. Smart, F. Guo, High quality low complexity scalable wavelet audio coding, in: Proceedings of IEEE International Conference Acoustics, Speech, and Signal Processing (ICASSP'97), Apr 1997, pp. 327–330.

N.E. Huang, Z. Shen, S.R. Long, M.C. Wu, H.H. Shih, Q. Zheng, et al., The empirical mode decomposition and the Hilbert spectrum for nonlinear and non-stationary time series analysis, Proc. R. Soc. Lond. A Math. Phys. Sci. 454 (1971) 903–995.

B. Kotnik, Z. Kacic, B. Horvat, The usage of wavelet packet transformation in automatic noisy speech recognition systems, in: IEEE, 2, 2003, pp. 131–134.

C.C.E. de Abreu, M.A.Q. Duarte, F. Villarreal, An immunological approach based on the negative selection algorithm for real noise classification in speech signals, AEU-Int. J. Electron. C 72 (2017) 125–133.

R. Gomez, T. Kawahara, K. Nakadai, Optimized wavelet-domain filtering under noisy and reverberant conditions, APSIPA Trans. Signal Inf. Process. 4 (e3) (2015) 1–12.

A. Adiga, M. Magimai, C.S. Seelamantula, Gammatone wavelet cepstral coefficients for robust speech recognition, in: 2013 IEEE International Conference of IEEE Region 10 (TENCON 2013), Xi'an, 2013, pp. 1–4.

Z. Xueying, J. Zhiping, Speech recognition based on auditory wavelet packet filter, in: Proc. 7th Int. Conf. Signal Process. 2004 (ICSP '04), vol 1, 2004, pp. 695–698.

O. Farooq, S. Datta, Mel filter-like admissible wavelet packet structure for speech recognition, IEEE Signal Process. Lett. 8 (7) (2001) 196–198.

J.N. Gowdy, Z. Tufekci, Mel-scaled discrete wavelet coefficients for speech recognition, in: Proceedings of the IEEE International Conference on Acoustics, Speech, and Signal Processing (ICASSP 2000), vol. 3, 2000, pp. 1351–1354.

A. Bandini, et al., Markerless analysis of articulatory movements in patients with Parkinson's disease, J. Voice 30 (6) (2016) 766-e1.

M. Nilashi, et al., A hybrid intelligent system for the prediction of Parkinson's disease progression using machine learning techniques, Biocybern. Biomed. Eng. 38 (1) (2018) 1–15.

S. Skodda, W. Visser, U. Schlegel, Vowel articulation in Parkinson's disease, J. Voice 25 (4) (2011) 467–472.

Chapter 10

A Framework for Artificially Intelligent Customized Voice Response System Design using Speech Synthesis Markup Language

Sajal Saha, Mazid Alam and Smita Dey
Department of CSE, Kaziranga University, Jorhat, India

10.1 Introduction

The idea of the Internet of Things (IoT) [1–5] came into the picture to control and monitor the connected devices over the Internet, which is increasing day by day. Earlier IoT enabled devices were controlled and monitored by smartphone apps, which are platform dependent (Android, iOS, Windows). An artificially intelligent voice response system is a better solution in this context to control the IoT devices. The primary objective in developing a voice response system is to build a human-computer interaction system that enables the system to recognize the human voice in the form of instruction. Every speech recognition system [6–8] filters out noises and identifies human voices, especially one that uses its wake word. Each voice response system has its own wake word. Amazon Echo's [9] wake word is "*Alexa*," which can be replaced by "*Echo*" or "*Amazon*." Google Home is equipped with wake word, "*Hey Google*" or "*Ok Google*." Apple HomePod is equipped with wake word "*Hey Siri*," etc. The wake word is a single word [10–12] or a phrase that when uttered, the voice response system circuits are activated and reject all other words, phrases, noises and other external acoustic events with virtually 100 percent accuracy. Alexa Voice Service and Google Assistant recognize and respond only to the designated user apart from the wake word. This feature provides added security to the voice response system.

One of the key limitations in the available voice response systems (Amazon Echo, Google Home, Apple HomePod) is that they respond to generic queries.

Intelligent Speech Signal Processing. https://doi.org/10.1016/B978-0-12-818130-0.00010-6

This limitation motivated us to design and develop a customized artificially intelligent, low-cost voice response system that responds to user-centered queries.

The rest of the chapter is organized as follows: Section 10.2 summarizes the survey of the related work done by the researchers worldwide. We provide an overview of the challenges faced by the app-controlled, IoT enabled devices and how the voice-controlled, IoT enabled devices are being introduced to overcome those challenges. We summarize the pros and cons of the available AI-based voice response systems for background information.

Sections 10.3–10.5 introduce the basics of the AWS IoT paradigm and its components: Alexa voice service (AVS) and AWS Lambda. Section 10.6 describes how message passing is done in a secure communication channel through Message Queuing Telemetry Transport (MQTT) protocol in our system. Section 10.7 describes the proposed architecture of the system and its workflow. Section 10.8 concludes the chapter.

10.2 Literature Survey

The author [13] connected home appliances such as lights and fans with AWS IoT and controlled it through Alexa Voice Service. These devices require frequent firmware updates, which are being done through Node-RED, an IBM developed visual wiring tool. The author uses Raspberry Pi and Intel Edison Board as the Alexa voice client device. Veton et al. [14, 15] have made a comparative analysis of various available speech recognition systems like Microsoft API, CMUSphinx, and Google API. Sphinx 4 is an open source speech recognition system developed by Carnegie Mellon University. Authors have used audio recordings as input and calculated the word error rate to make the comparison, and found that Google API outperformed the other speech recognition systems. Authors excluded Alexa voice service from the test. In our test, we saw that Alexa voice service performed better than Google API in a noisy environment. The authors [14] designed a virtual personal assistant system for human-computer interaction in a multimodal environment where given inputs are speech, image, video, manual gesture, gaze, and body movement. IBM Watson Analytics also provides a development environment like Alexa voice service. IBM Watson provides speech to text, text to speech, conversation, tone analyzer, and visual recognition services to the user. The authors [14] have used IBM Watson with Node-RED and Python to connect the IoT enabled devices through API.

10.3 AWS IoT

Amazon Web Services Internet of Things (AWS IoT) is a cloud-based Platform as a Service (PAAS) that enables users to connect, control, and monitor IoT devices from a remote location. It is a subscription-based service. Users are charged based on the number of devices enabled in AWS IoT. The user can easily enable/disable a device from the AWS management console. AWS IoT provides Amazon voice services (AVS) to the Amazon voice client.

10.4 Amazon Voice Service (AVS)

AVS helps the developer to connect voice-enabled IoT devices that have a microphone connectivity. We call this voice-enabled device Alexa Voice Client. After the integration of AVS with Alexa Voice Client, it will have access all the relevant API's that Alexa voice client receives as an instruction from the user. Middleware APIs convert the request to a response. AVS has the access to the entire built-in API available in AWS IoT. The users can build their own API uploading Alexa Skills Kit in AWS IoT to make the voice response system (Alexa voice client) feature user-friendly and able to reach. AVS creates an interface with Alexa voice client and enables three functionalities: speech recognition, voice control, and audio playback. Each interface function is driven by two types of messages: either through message directives or message events. Directives are the messages sent from the cloud to take action, and events are messages sent by the client device to cloud to show what action is being taken. Directives and event messages are generated through AWS Lambda, speech synthesis markup language, and Alexa skills kit.

10.5 AWS Lambda

AWS Lambda is another PaaS services that enable stateless development environment to the user. The user writes the piece of code in AWS Lambda function to control the connected device. Alexa voice service recognizes the speech recognition from Alexa voice client and converts it into an input to AWS Lambda function. Speech recognition to input instruction conversion is done through an API SpeechRecognizer.Recognize available in Alexa voice service as shown in Fig. 10.1. AWS Lambda function processes the input and generates customized output in the form of event messages as shown in Fig. 10.1. Event messages synthesize speech signals through another API called SpeechSynthesizer.Speak as shown in Fig. 10.1.

10.6 Message Queuing Telemetry Transport (MQTT)

Message Queuing Telemetry Transport (MQTT) is a message passing protocol [16, 17] like TCP and UDP. MQTT uses low bandwidth and low latency during message transmission. It is a machine to machine communication protocol. There are three components in MQTT: publisher, broker, and subscriber. The subscriber/user sends a request to the publisher via the broker as shown in Fig. 10.2. The publisher responds to the request from the subscriber via the broker in a secure communication channel. The broker acts like a filter allowing

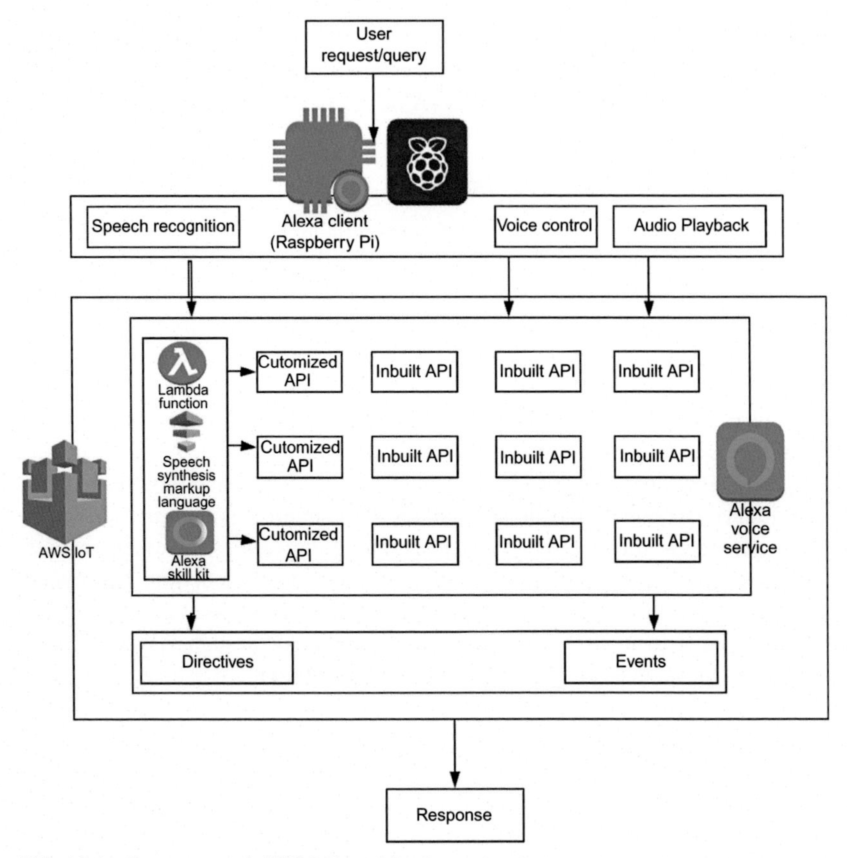

FIG. 10.1 Components of AWS IoT used for the proposed system.

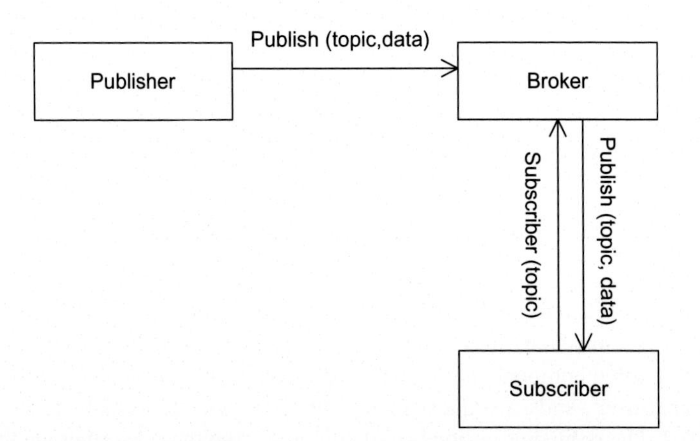

FIG. 10.2 Components of MQTT.

only those data requested by the subscriber. We have implemented MQTT configuration using Eclipse Paho. Eclipse Paho is an open source MQTT client implementation platform. The Paho Java Client is an MQTT client library written in Java for developing applications that run on the JVM or other Java compatible platforms such as Android.

10.7 Proposed Architecture

We have designed the customized voice response system using Raspberry Pi as Alexa voice client. Raspberry Pi will be used as an MQTT client, and for message passing between the user and Alexa voice service (AVS). The hardware and software requirement are shown in Tables 10.1 and 10.2, respectively.

The architecture of the proposed system is shown in Fig. 10.3. We have used the relay switch to enable home appliances such as lights and fans. We connect the relay module to the Raspberry Pi using jumper wires. We use the following gpio pins to connect the relay to our Pi:

For the kitchen light, Pin No. 23 is used (P1 header pin 16),
For the bedroom light, Pin No. 10 is used (P1 header pin 19),
For the bedroom fan, Pin No. 11 is used (P1 header pin 23), and
For the bathroom light, Pin No. 22 is used (P1 header pin 15).

We have used gpio pins for power supply and ground, respectively. Also, we have used a button to connect to the Pi so we can activate Alexa by pressing the button. We can alternatively activate Alexa by even giving a wake word signal.

Fig. 10.4 gives a brief overview of the system. The entire workflow of the system is shown in a flowchart in Fig. 10.5.

The user sends a voice request to Raspberry Pi. Raspberry Pi acts here as Alexa voice client. Raspberry Pi transmits the speech signal to Alexa voice service (AVS). AVS eliminates noise from the speech signal and converts it to text. The text is synthesized as an instruction that invokes the Alexa skills kit, which

TABLE 10.1 Hardware Requirements

S. No	Material	Quantity
1	Raspberry Pi 3 model B+	1
2	Relay (generic)	4
3	Jumper wires	30
4	Soundcard or microphone	1
5	Speakers	1

TABLE 10.2 Software Requirements

S. No	Software Applications and Online Services
1	Amazon Alexa/Alexa skills kit
2	Amazon Alexa/Alexa voice service
3	Amazon web services/AWS IoT
4	Amazon web services/AWS Lambda

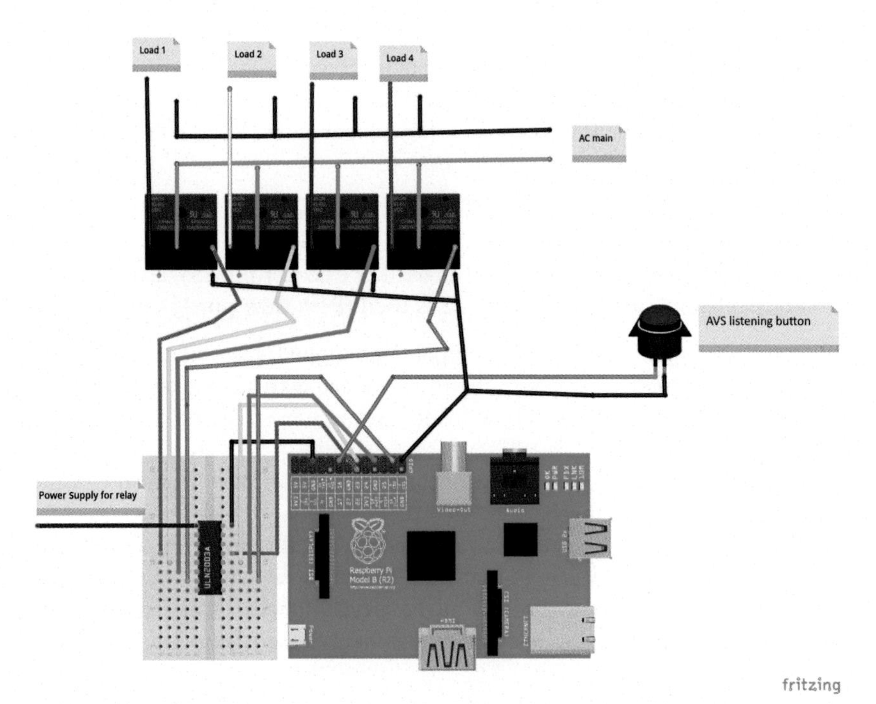

FIG. 10.3 System architecture of the proposed voice response system.

will accept user command, and then searches for the suitable API using AWS Lambda. Few built-in APIs are present in Alexa skills kit.

When the voice response system is asked a query, Alexa voice service invokes three APIs namely:

a. SpeechRecognizer.Recognize
b. Speech Synthesizer.Speech Started
c. Speech Synthesizer.Speak

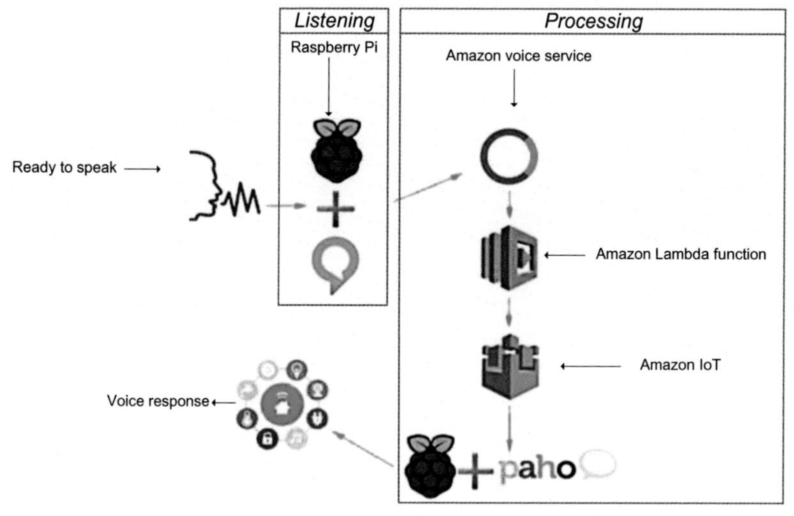

FIG. 10.4 Overview of the proposed system.

For example,

User: "Alexa, How are you?"
Alexa: "I am feeling like the top of the world."
Invoked API's code are as follows:
API: SpeechRecognizer.Recognize


```
{
 "event": {
   "header": {
     "dialogRequestId": "dialogRequestId-7b155828-6eb7-4859-870f-ca7431bcd66d",
     "namespace": "SpeechRecognizer",
     "name": "Recognize",
     "messageId": "b28c0283-2a3b-420d-98ab-5833196fa90d"
   },
   "payload": {
     "profile": "CLOSE_TALK",
     "format": "AUDIO_L16_RATE_16000_CHANNELS_1"
   }
 },
 "context": [
   {
     "header": {
       "namespace": "AudioPlayer",
       "name": "PlaybackState"
     },
```

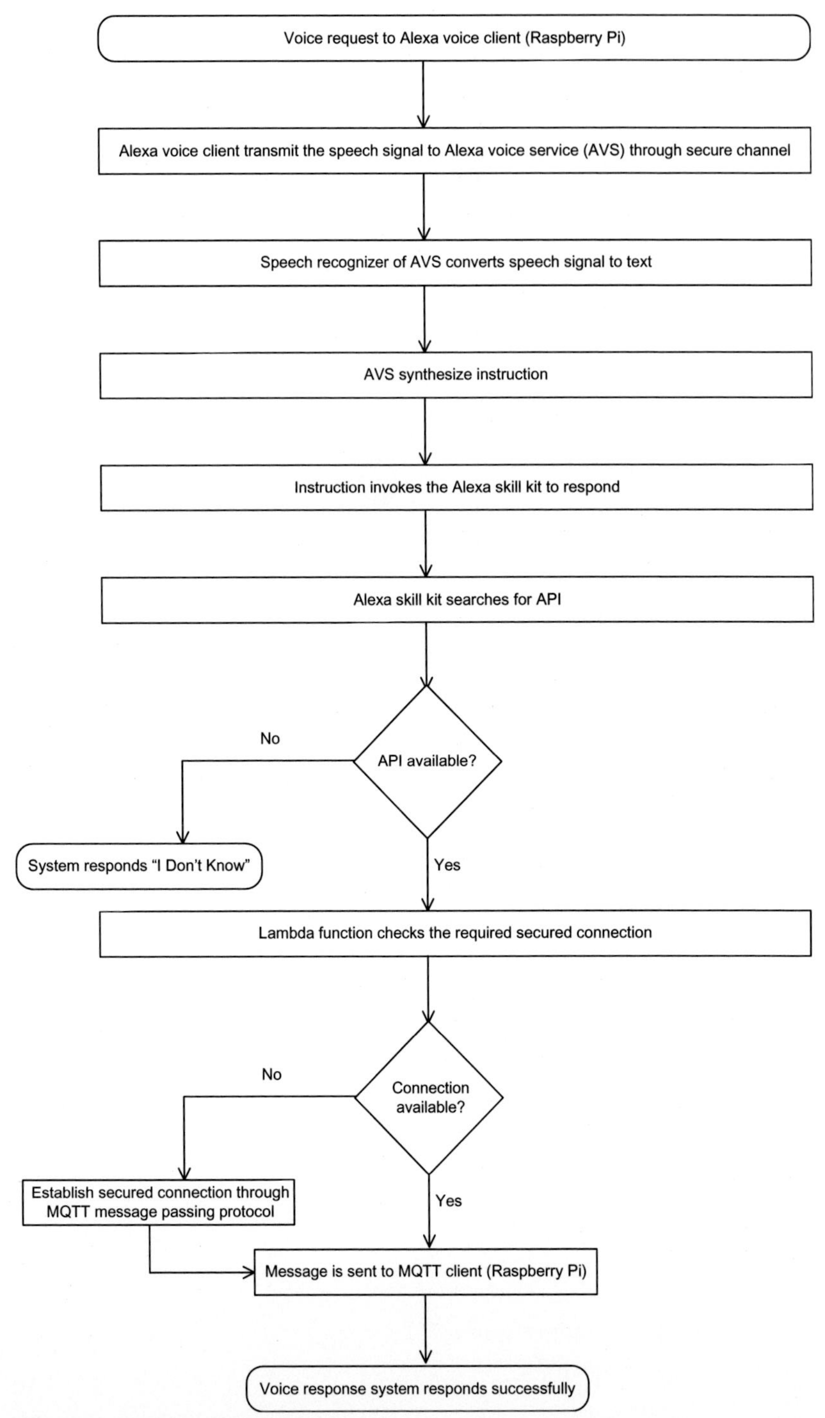

FIG. 10.5 Workflow of the proposed system.

```
        "payload": {
          "token": "",
          "playerActivity": "IDLE",
          "offsetInMilliseconds": "0"
        }
      }
    ]
  }
```


API: Speech Synthesizer.Speech Started


```
{
  "event": {
    "header": {
      "namespace": "SpeechSynthesizer",
      "name": "SpeechStarted",
      "messageId": "8e18c39a-f900-47ca-bc70-0686680ba902"
    },
    "payload": {
      "token": "amzn1.as-ct.v1.Domain:Application:Knowledge#ACRI#
f42592cf-a445-4485-99c5-36cf36e18d17"
    }
  },
  "context": []
}
```


API: SpeechSynthesizer.Speak


```
{
  "directive": {
   "header": {
    "dialogRequestId": "dialogRequestId-7b155828-6eb7-4859-870f-ca7431bcd66d",
    "namespace": "SpeechSynthesizer",
    "name": "Speak",
    "messageId": "ad6a54bd-ad9c-4a23-ae56-9eb3da8d8dc8"
  },
  "payload": {
   "url": "cid:f42592cf-a445-4485-99c5-36cf36e18d17_1932182893",
   "token": "amzn1.as-ct.v1.Domain:Application:Knowledge#ACRI#
f42592cf-a445-4485-99c5-36cf36e18d17",
   "format": "AUDIO_MPEG"
  }
 }
}
```


If we identify that our system fails to respond to a certain query, we write our own instruction using AWS Lambda function and speech synthesis markup language

(SSML) and upload it as a customized API in Alexa skills kit. After the identification of a suitable API, the Lambda function requests AVS to check the existence of the required connection and sends the response to Raspberry Pi. Raspberry Pi now acts as the MQTT client and gives the response back to the user.

10.8 Conclusion

The human-computer interaction system has traversed a long path. It initiated its journey through keyboard-mouse, apps and, now, it is voice. Today, all electronic devices are connected through the Internet. These devices are controlled through apps that are platform dependent (Android, iOS, Windows) and requires frequent firmware updates. Our proposed system eliminates platform dependency, and it recognizes and responds only to the designated user. This feature provides added security to our voice response system. Moreover, this system costs around 3,400 INR ($42), which is cheaper than the available voice response systems in the market (Amazon Echo, Google Home, Apple HomePod).

It is a well-known controversy that Amazon Echo stores user's private conversations [18] in AWS cloud server, although Amazon denies the fact. Our proposed framework stores AVS collected data in an individual user's account. So, there is no risk of the user's data leakage or exposure.

During testing and validation of the proposed system, we observed the following: The proposed voice response system replies with 100% accuracy for a general query like, "Who is the prime minister of India," but fails to reply to a user-specific query like "Where is Dr. Sajal Saha right now?" We are working on integrating all the Google apps (except Gmail), such as Google Maps, Google Calendar, Google Find my Phone, and Google Classroom through custom API, so that proposed voice response system can give a precise answer to the user-specific query.

References

[1] M.G. Al-fuqaha, M. Mohammadi, M. Aledhari, M. Ayyash, Internet of things: a survey on enabling technologies, protocols and applications internet of things: a survey on enabling technologies, Protocols Appl. 17 (January) (2015) 2347–2376.

[2] C. Bhatt, N. Dey, A.S. Ashour (Eds.), Internet of Things and Big Data Technologies for Next Generation Healthcare, Springer, 2017.

[3] G. Elhayatmy, N. Dey, A.S. Ashour, Internet of Things and Big Data Analytics Toward Next-Generation Intelligence, vol. 30, Springer, 2018, pp. 3–20.

[4] N. Dey, A.S. Ashour, C. Bhatt, Internet of Things and Big Data Technologies for Next Generation Healthcare, vol. 23, Springer, 2017, pp. 3–12.

[5] N. Dey, A. Mukherjee, Embedded Systems and Robotics with Open Source Tools, CRC Press, 2016.

[6] N. Dey, A.S. Ashour, Applied Examples and Applications of Localization and Tracking Problem of Multiple Speech Sources, Springer, 2018, pp. 35–48.

[7] A. Singh, N. Dey, A.S. Ashour, Scope of automation in semantics-driven multimedia information retrieval from web, in: Web Semantics for Textual and Visual Information Retrieval, IGI Global, 2017, pp. 1–16.
[8] N. Dey, V. Santhi (Eds.), Intelligent Techniques in Signal Processing for Multimedia Security, Springer International Publishing, 2017.
[9] P. Dempsey, Amazon Echo digital personal assistant, No. March, 2015, pp. 88–89.
[10] V. Këpuska, Wake-Up-Word Speech Recognition, No. June 2011, 2014.
[11] V.Z. Këpuska, T.B. Klein, Nonlinear analysis a novel wake-up-word speech recognition system, wake-up-word recognition task, technology and evaluation, Nonlinear Anal. 71 (12) (2009) e2772–e2789.
[12] V. Këpuska, C. Engineering, Wake-up-word speech recognition application for first responder communication enhancement, in: Proceedings Volume 6201, Sensors, and Command, Control, Communications, and Intelligence (C3I) Technologies for Homeland Security and Homeland Defense V, 2006, pp. 1–8.
[13] A. Rajalakshmi, H. Shahnasser, Internet of things using node-red and alexa, in: 2017 17th Int. Symp. Commun. Inf. Technol. Isc. 2017, vol. 2018–Janua, 2018, pp. 1–4.
[14] V. Kepuska, G. Bohouta, Next-generation of virtual personal assistants (Microsoft Cortana, Apple Siri, Amazon Alexa and Google Home), in: 2018 IEEE 8th Annu. Comput. Commun. Work. Conf. CCWC 2018, vol. 2018–Janua, no. c, 2018, pp. 99–103.
[15] V. Këpuska, Comparing speech recognition systems (Microsoft API, Google API And CMU Sphinx), Int. J. Eng. Res. Appl. 07 (03) (2017) 20–24.
[16] G. C. Hillar, MQTT Essentials—A Lightweight IoT Protocol, 1st ed., Pact.
[17] Y. Upadhyay, MQTT Based Secured Home Automation System, 2016.
[18] H. Chung, M. Iorga, and J. Voas, Alexa, Can I Trust You?.

Index

Note: Page numbers followed by *f* indicate figures and *t* indicate tables.

K

L

M

N

O

P

R

S

W

Printed in the United States
By Bookmasters